A FIRST

LOOK

AT
RIGOROUS
PROBABILITY
THEORY

Second Edition

A FIRST

LOOK

AT
RIGOROUS
PROBABILITY
THEORY

Second Edition

Jeffrey S. Rosenthal

University of Toronto, Canada

World Scientific

NEW JERSEY • LONDON • SINGAPORE • BEIJING • SHANGHAI • HONG KONG • TAIPEI • CHENNAI

Published by

World Scientific Publishing Co. Pte. Ltd.
5 Toh Tuck Link, Singapore 596224
USA office: 27 Warren Street, Suite 401-402, Hackensack, NJ 07601
UK office: 57 Shelton Street, Covent Garden, London WC2H 9HE

Library of Congress Cataloging-in-Publication Data
Rosenthal, Jeffrey S. (Jeffrey Seth)
  A first look at rigorous probability theory, 2nd ed. / Jeffrey S. Rosenthal.
  p. cm.
  Includes bibliographical references and index.
  ISBN-13 978-981-270-370-5
  ISBN-10 981-270-370-5
  ISBN-13 978-981-270-371-2 (pbk)
  ISBN-10 981-270-371-3 (pbk)
  1. Probabilites. 2. Measure theory. 3. Probability measures.
  QA273 .R7835 2006
  519.2--dc22
                                                    2007280482

British Library Cataloguing-in-Publication Data
A catalogue record for this book is available from the British Library.

First published 2006
Reprinted 2008, 2009, 2010, 2011

Printed in Singapore.

To my wonderful wife, Margaret Fulford:

always supportive, always encouraging.

Preface to the First Edition.

This text grew out of my lecturing the Graduate Probability sequence STA 2111F / 2211S at the University of Toronto over a period of several years. During this time, it became clear to me that there are a large number of graduate students from a variety of departments (mathematics, statistics, economics, management, finance, computer science, engineering, etc.) who require a working knowledge of rigorous probability, but whose mathematical background may be insufficient to dive straight into advanced texts on the subject.

This text is intended to answer that need. It provides an introduction to rigorous (i.e., mathematically precise) probability theory using measure theory. At the same time, I have tried to make it brief and to the point, and as accessible as possible. In particular, probabilistic language and perspective are used throughout, with necessary measure theory introduced only as needed.

I have tried to strike an appropriate balance between rigorously covering the subject, and avoiding unnecessary detail. The text provides mathematically complete proofs of all of the essential introductory results of probability theory and measure theory. However, more advanced and specialised areas are ignored entirely or only briefly hinted at. For example, the text includes a complete proof of the classical Central Limit Theorem, including the necessary Continuity Theorem for characteristic functions. However, the Lindeberg Central Limit Theorem and Martingale Central Limit Theorem are only briefly sketched and are not proved. Similarly, all necessary facts from measure theory are proved before they are used. However, more abstract and advanced measure theory results are not included. Furthermore, the measure theory is almost always discussed purely in terms of probability, as opposed to being treated as a separate subject which must be mastered before probability theory can be studied.

I hesitated to bring these notes to publication. There are many other books available which treat probability theory with measure theory, and some of them are excellent. For a partial list see Subsection B.3 on page 210. (Indeed, the book by Billingsley was the textbook from which I taught before I started writing these notes. While much has changed since then, the knowledgeable reader will still notice Billingsley's influence in the treatment of many topics herein. The Billingsley book remains one of the best sources for a complete, advanced, and technically precise treatment of probability theory with measure theory.) In terms of content, therefore, the current text adds very little indeed to what has already been written. It was only the reaction of certain students, who found the subject easier to learn from my notes than from longer, more advanced, and more all-inclusive books, that convinced me to go ahead and publish. The reader is urged to consult other books for further study and additional detail.

There are also many books available (see Subsection B.2) which treat probability theory at the undergraduate, less rigorous level, without the use of general measure theory. Such texts provide intuitive notions of probabilities, random variables, etc., but without mathematical precision. In this text it will generally be assumed, for purposes of intuition, that the student has at least a passing familiarity with probability theory at this level. Indeed, Section 1 of the text attempts to link such intuition with the mathematical precision to come. However, mathematically speaking we will not require many results from undergraduate-level probability theory.

Structure. The first six sections of this book could be considered to form a "core" of essential material. After learning them, the student will have a precise mathematical understanding of probabilities and σ-algebras; random variables, distributions, and expected values; and inequalities and laws of large numbers. Sections 7 and 8 then diverge into the theory of gambling games and Markov chain theory. Section 9 provides a bridge to the more advanced topics of Sections 10 through 14, including weak convergence, characteristic functions, the Central Limit Theorem, Lebesgue Decomposition, conditioning, and martingales.

The final section, Section 15, provides a wide-ranging and somewhat less rigorous introduction to the subject of general stochastic processes. It leads up to diffusions, Itô's Lemma, and finally a brief look at the famous Black-Scholes equation from mathematical finance. It is hoped that this final section will inspire readers to learn more about various aspects of stochastic processes.

Appendix A contains basic facts from elementary mathematics. This appendix can be used for review and to gauge the book's level. In addition, the text makes frequent reference to Appendix A, especially in the earlier sections, to ease the transition to the required mathematical level for the subject. It is hoped that readers can use familiar topics from Appendix A as a springboard to less familiar topics in the text.

Finally, Appendix B lists a variety of references, for background and for further reading.

Exercises. The text contains a number of exercises. Those very closely related to textual material are inserted at the appropriate place. Additional exercises are found at the end of each section, in a separate subsection. I have tried to make the exercises thought provoking without being too difficult. Hints are provided where appropriate. Rather than always asking for computations or proofs, the exercises sometimes ask for explanations and/or examples, to hopefully clarify the subject matter in the student's mind.

Prerequisites. As a prerequisite to reading this text, the student should have a solid background in basic undergraduate-level real analysis (*not* including measure theory). In particular, the mathematical background summarised in Appendix A should be very familiar. If it is not, then books such as those in Subsection B.1 should be studied first. It is also helpful, but not essential, to have seen some undergraduate-level probability theory at the level of the books in Subsection B.2.

Further reading. For further reading beyond this text, the reader should examine the similar but more advanced books of Subsection B.3. To learn additional topics, the reader should consult the books on pure measure theory of Subsection B.4, and/or the advanced books on stochastic processes of Subsection B.5, and/or the books on mathematical finance of Subsection B.6. I would be content to learn only that this text has inspired students to look at more advanced treatments of the subject.

Acknowledgements. I would like to thank several colleagues for encouraging me in this direction, in particular Mike Evans, Andrey Feuerverger, Keith Knight, Omiros Papaspiliopoulos, Jeremy Quastel, Nancy Reid, and Gareth Roberts. Most importantly, I would like to thank the many students who have studied these topics with me; their questions, insights, and difficulties have been my main source of inspiration.

Jeffrey S. Rosenthal
Toronto, Canada, 2000
jeff@math.toronto.edu
http://probability.ca/jeff/

Second Printing (2003). For the second printing, a number of minor errors have been corrected. Thanks to Tom Baird, Meng Du, Avery Fullerton, Longhai Li, Hadas Moshonov, Nataliya Portman, and Idan Regev for helping to find them.

Third Printing (2005). A few more minor errors were corrected, with thanks to Samuel Hikspoors, Bin Li, Mahdi Lotfinezhad, Ben Reason, Jay Sheldon, and Zemei Yang.

Preface to the Second Edition.

I am pleased to have the opportunity to publish a second edition of this book. The book's basic structure and content are unchanged; in particular, the emphasis on establishing probability theory's rigorous mathematical foundations, while minimising technicalities as much as possible, remains paramount. However, having taught from this book for several years, I have made considerable revisions and improvements. For example:

- Many small additional topics have been added, and existing topics expanded. As a result, the second edition is over forty pages longer than the first.
- Many new exercises have been added, and some of the existing exercises have been improved or "cleaned up". There are now about 275 exercises in total (as compared with 150 in the first edition), ranging in difficulty from quite easy to fairly challenging, many with hints provided.
- Further details and explanations have been added in steps of proofs which previously caused confusion.
- Several of the longer proofs are now broken up into a number of lemmas, to more easily keep track of the different steps involved, and to allow for the possibility of skipping the most technical bits while retaining the proof's overall structure.
- A few proofs, which are required for mathematical completeness but which require advanced mathematics background and/or add little understanding, are now marked as "optional".
- Various interesting, but technical and inessential, results are presented as remarks or footnotes, to add information and context without interrupting the text's flow.
- The Extension Theorem now allows the original set function to be defined on a semialgebra rather than an algebra, thus simplifying its application and increasing understanding.
- Many minor edits and rewrites were made throughout the book to improve the clarity, accuracy, and readability.

I thank Ying Oi Chiew and Lai Fun Kwong of World Scientific for facilitating this edition, and thank Richard Dudley, Eung Jun Lee, Neal Madras, Peter Rosenthal, Hermann Thorisson, and Bálint Virág for helpful comments. Also, I again thank the many students who have studied and discussed these topics with me over many years.

Jeffrey S. Rosenthal
Toronto, Canada, 2006

Second Printing (2007). A few very minor corrections were made, with thanks to Joe Blitzstein, Saad Siddiqui, and Emil Zeuthen.

Third Printing (2009). No changes were made.

Fourth Printing (2010). Some additional small errors were corrected, with thanks to Örn Arnaldsson, Bent Jørgensen, Chris Mansley, Kohei Nagamachi, Patrick Rabau, Mohsen Soltanifar, and Hermann Thorisson.

Fifth Printing (2011). A few more small corrections were made, with thanks to David Alexander, Martin Hazelton, Andrea Lecchini-Visintini, Gareth Roberts, Igal Sason, Mohsen Soltanifar, and Albert Zevelev.

Contents

1. The need for measure theory.

This introductory section is directed primarily to those readers who have some familiarity with undergraduate-level probability theory, and who may be unclear as to why it is necessary to introduce measure theory and other mathematical difficulties in order to study probability theory in a rigorous manner.

We attempt to illustrate the limitations of undergraduate-level probability theory in two ways: the restrictions on the kinds of random variables it allows, and the question of what sets can have probabilities defined on them.

1.1. Various kinds of random variables.

The reader familiar with undergraduate-level probability will be comfortable with a statement like, "Let X be a random variable which has the **Poisson**(5) distribution." The reader will know that this means that X takes as its value a "random" non-negative integer, such that the integer $k \geq 0$ is chosen with probability $\mathbf{P}(X = k) = e^{-5}5^k/k!$. The expected value of, say, X^2, can then be computed as $\mathbf{E}(X^2) = \sum_{k=0}^{\infty} k^2 e^{-5}5^k/k!$. X is an example of a *discrete random variable*.

Similarly, the reader will be familiar with a statement like, "Let Y be a random variable which has the **Normal**(0, 1) distribution." This means that the probability that Y lies between two real numbers $a < b$ is given by the integral $\mathbf{P}(a \leq Y \leq b) = \int_a^b \frac{1}{\sqrt{2\pi}} e^{-y^2/2} dy$. (On the other hand, $\mathbf{P}(Y = y) = 0$ for any particular real number y.) The expected value of, say, Y^2, can then be computed as $\mathbf{E}(Y^2) = \int_{-\infty}^{\infty} y^2 \frac{1}{\sqrt{2\pi}} e^{-y^2/2} dy$. Y is an example of an *absolutely continuous random variable*.

But now suppose we introduce a new random variable Z, as follows. We let X and Y be as above, and then flip an (independent) fair coin. If the coin comes up heads we set $Z = X$, while if it comes up tails we set $Z = Y$. In symbols, $\mathbf{P}(Z = X) = \mathbf{P}(Z = Y) = 1/2$. Then what sort of random variable is Z? It is not discrete, since it can take on an uncountable number of different values. But it is not absolutely continuous, since for certain values z (specifically, when z is a non-negative integer) we have $\mathbf{P}(Z = z) > 0$. So how can we study the random variable Z? How could we compute, say, the expected value of Z^2?

The correct response to this question, of course, is that the division of random variables into discrete versus absolutely continuous is artificial. Instead, measure theory allows us to give a common definition of expected value, which applies equally well to discrete random variables (like X above), to continuous random variables (like Y above), to combinations of them (like

Z above), and to other kinds of random variables not yet imagined. These issues are considered in Sections 4, 6, and 12.

1.2. The uniform distribution and non-measurable sets.

In undergraduate-level probability, continuous random variables are often studied in detail. However, a closer examination suggests that perhaps such random variables are not completely understood after all.

To take the simplest case, suppose that X is a random variable which has the uniform distribution on the unit interval $[0, 1]$. In symbols, $X \sim$ **Uniform**$[0, 1]$. What precisely does this mean?

Well, certainly this means that $\mathbf{P}(0 \le X \le 1) = 1$ It also means that $\mathbf{P}(0 \le X \le 1/2) = 1/2$, that $\mathbf{P}(3/4 \le X \le 7/8) = 1/8$, etc., and in general that $\mathbf{P}(a \le X \le b) = b - a$ whenever $0 \le a \le b \le 1$, with the same formula holding if $\le$ is replaced by $<$. We can write this as

$$\mathbf{P}([a, b]) = \mathbf{P}((a, b]) = \mathbf{P}([a, b)) = \mathbf{P}((a, b)) = b - a, \quad 0 \le a \le b \le 1. \tag{1.2.1}$$

In words, the probability that X lies in any interval contained in $[0, 1]$ is simply the length of the interval. (We include in this the degenerate case when $a = b$, so that $\mathbf{P}(\{a\}) = 0$ for the singleton set $\{a\}$; in words, the probability that X is equal to any particular number a is zero.)

Similarly, this means that

$$\mathbf{P}(1/4 \le X \le 1/2 \underline{\text{ or }} 2/3 \le X \le 5/6)$$

$$= \mathbf{P}(1/4 \le X \le 1/2) + \mathbf{P}(2/3 \le X \le 5/6) = 1/4 + 1/6 = 5/12,$$

and in general that if A and B are disjoint subsets of $[0, 1]$ (for example, if $A = [1/4, \ 1/2]$ and $B = [2/3, \ 5/6]$), then

$$\mathbf{P}(A \cup B) = \mathbf{P}(A) + \mathbf{P}(B). \tag{1.2.2}$$

Equation (1.2.2) is called *finite additivity*.

Indeed, to allow for countable operations (such as limits, which are extremely important in probability theory), we would like to extend (1.2.2) to the case of a countably infinite number of disjoint subsets: if $A_1, A_2, A_3, \ldots$ are disjoint subsets of $[0, 1]$, then

$$\mathbf{P}(A_1 \cup A_2 \cup A_3 \cup \ldots) = \mathbf{P}(A_1) + \mathbf{P}(A_2) + \mathbf{P}(A_3) + \ldots. \tag{1.2.3}$$

Equation (1.2.3) is called *countable additivity*.

Note that we do *not* extend equation (1.2.3) to *uncountable* additivity. Indeed, if we did, then we would expect that $\mathbf{P}([0, 1]) = \sum_{x \in [0,1]} \mathbf{P}(\{x\})$, which is clearly false since the left-hand side equals 1 while the right-hand side equals 0. (There is no contradiction to (1.2.3) since the interval $[0, 1]$ is not countable.) It is for this reason that we restrict attention to countable operations. (For a review of countable and uncountable sets, see Subsection A.2. Also, recall that for non-negative uncountable collections $\{r_\alpha\}_{\alpha \in I}$, $\sum_{\alpha \in I} r_\alpha$ is defined to be the supremum of $\sum_{\alpha \in J} r_\alpha$ over finite $J \subseteq I$.)

Similarly, to reflect the fact that X is "uniform" on the interval $[0, 1]$, the probability that X lies in some subset should be unaffected by "shifting" (with wrap-around) the subset by a fixed amount. That is, if for each subset $A \subseteq [0, 1]$ we define the r-shift of A by

$$A \oplus r \equiv \{a + r \,;\, a \in A,\ a + r \leq 1\} \cup \{a + r - 1 \,;\, a \in A,\ a + r > 1\}, \quad (1.2.4)$$

then we should have

$$\mathbf{P}(A \oplus r) = \mathbf{P}(A), \qquad 0 \leq r \leq 1. \tag{1.2.5}$$

So far so good. But now suppose we ask, what is the probability that X is rational? What is the probability that X^n is rational for some positive integer n? What is the probability that X is *algebraic*, i.e. the solution to some polynomial equation with integer coefficients? Can we compute these things? More fundamentally, are all probabilities such as these necessarily even *defined*? That is, does $\mathbf{P}(A)$ (i.e., the probability that X lies in the subset A) even make *sense* for every possible subset $A \subseteq [0, 1]$?

It turns out that the answer to this last question is no, as the following proposition shows. The proof requires equivalence relations, but can be skipped if desired since the result is not used elsewhere in this book.

Proposition 1.2.6.   *There does not exist a definition of $\mathbf{P}(A)$, defined for all subsets $A \subseteq [0, 1]$, satisfying (1.2.1) and (1.2.3) and (1.2.5).*

Proof (optional).   Suppose, to the contrary, that $\mathbf{P}(A)$ could be so defined for each subset $A \subseteq [0, 1]$. We will derive a contradiction to this.

Define an equivalence relation (see Subsection A.5) on $[0, 1]$ by: $x \sim y$ if and only if the difference $y - x$ is rational. This relation partitions the interval $[0, 1]$ into a disjoint union of equivalence classes. Let H be a subset of $[0, 1]$ consisting of precisely one element from each equivalence class (such H must exist by the Axiom of Choice, see page 200). For definiteness, assume that $0 \notin H$ (say, if $0 \in H$, then replace it by $1/2$).

Now, since H contains an element of each equivalence class, we see that each point in $(0, 1]$ is contained in the union $\bigcup_{\substack{r \in [0,1) \\ r \text{ rational}}} (H \oplus r)$ of shifts of H. Furthermore, since H contains just *one* point from each equivalence class, we see that these sets $H \oplus r$, for rational $r \in [0, 1)$, are all disjoint.

But then, by countable additivity (1.2.3), we have

$$\mathbf{P}((0,1]) = \sum_{\substack{r \in [0,1) \\ r \text{ rational}}} \mathbf{P}(H \oplus r).$$

Shift-invariance (1.2.5) implies that $\mathbf{P}(H \oplus r) = \mathbf{P}(H)$, whence

$$1 = \mathbf{P}((0,1]) = \sum_{\substack{r \in [0,1) \\ r \text{ rational}}} \mathbf{P}(H).$$

This leads to the desired contradiction: A countably infinite sum of the same quantity repeated can only equal 0, or ∞, or $-\infty$, but it can never equal 1. ∎

This proposition says that if we want our probabilities to satisfy reasonable[*] properties, then we *cannot* define them for all possible subsets of $[0,1]$. Rather, we must *restrict* their definition to certain "measurable" sets. This is the motivation for the next section.

Remark.   The existence of problematic sets like H above turns out to be *equivalent* to the Axiom of Choice. In particular, we can never define such sets explicitly – only implicitly via the Axiom of Choice as in the above proof.

1.3. Exercises.

Exercise 1.3.1.   Suppose that $\Omega = \{1,2\}$, with $\mathbf{P}(\emptyset) = 0$ and $\mathbf{P}\{1,2\} = 1$. Suppose $\mathbf{P}\{1\} = \frac{1}{4}$. Prove that $\mathbf{P}$ is countably additive if and only if $\mathbf{P}\{2\} = \frac{3}{4}$.

Exercise 1.3.2.   Suppose $\Omega = \{1,2,3\}$ and $\mathcal{F}$ is the collection of all subsets of Ω. Find (with proof) necessary and sufficient conditions on the real numbers x, y, and z, such that there exists a countably additive probability measure $\mathbf{P}$ on $\mathcal{F}$, with $x = \mathbf{P}\{1,2\}$, $y = \mathbf{P}\{2,3\}$, and $z = \mathbf{P}\{1,3\}$.

Exercise 1.3.3.   Suppose that $\Omega = \mathbf{N}$ is the set of positive integers, and $\mathbf{P}$ is defined for all $A \subseteq \Omega$ by $\mathbf{P}(A) = 0$ if A is finite, and $\mathbf{P}(A) = 1$ if A is infinite. Is $\mathbf{P}$ finitely additive?

Exercise 1.3.4.   Suppose that $\Omega = \mathbf{N}$, and $\mathbf{P}$ is defined for all $A \subseteq \Omega$ by $\mathbf{P}(A) = |A|$ if A is finite (where $|A|$ is the number of elements in

---

[*]In fact, assuming the Continuum Hypothesis, Proposition 1.2.6 continues to hold if we require only (1.2.3) and that $0 < \mathbf{P}([0,1]) < \infty$ and $\mathbf{P}\{x\} = 0$ for all x; see e.g. Billingsley (1995, p. 46).

the subset A), and $\mathbf{P}(A) = \infty$ if A is infinite. This $\mathbf{P}$ is of course not a probability measure (in fact it is *counting measure*), however we can still ask the following. (By convention, $\infty + \infty = \infty$.)

(a) Is $\mathbf{P}$ finitely additive?

(b) Is $\mathbf{P}$ countably additive?

Exercise 1.3.5.   **(a)** In what step of the proof of Proposition 1.2.6 was (1.2.1) used?

(b) Give an example of a countably additive set function $\mathbf{P}$, defined on *all* subsets of $[0, 1]$, which satisfies (1.2.3) and (1.2.5), but not (1.2.1).

1.4. Section summary.

In this section, we have discussed why measure theory is necessary to develop a mathematical rigorous theory of probability. We have discussed basic properties of probability measures such as additivity. We have considered the possibility of random variables which are neither absolutely continuous nor discrete, and therefore do not fit easily into undergraduate-level understanding of probability. Finally, we have proved that, for the uniform distribution on $[0, 1]$, it will not be possible to define a probability on every single subset.

2. Probability triples.

In this section we consider probability triples and how to construct them. In light of the previous section, we see that to study probability theory properly, it will be necessary to keep track of *which* subsets A have a probability $\mathbf{P}(A)$ defined for them.

2.1. Basic definition.

We define a *probability triple* or *(probability) measure space* or *probability space* to be a triple $(\Omega, \mathcal{F}, \mathbf{P})$, where:
- the *sample space* Ω is any non-empty set (e.g. $\Omega = [0, 1]$ for the uniform distribution considered above);
- the *σ-algebra* (read "sigma-algebra") or *σ-field* (read "sigma-field") $\mathcal{F}$ is a collection of subsets of Ω, containing Ω itself and the empty set $\emptyset$, and closed under the formation of complements[*] and countable unions and countable intersections (e.g. for the uniform distribution considered above, $\mathcal{F}$ would certainly contain all the intervals $[a, b]$, but would contain many more subsets besides);
- the *probability measure* $\mathbf{P}$ is a mapping from $\mathcal{F}$ to $[0, 1]$, with $\mathbf{P}(\emptyset) = 0$ and $\mathbf{P}(\Omega) = 1$, such that $\mathbf{P}$ is countably additive as in (1.2.3).

This definition will be in constant use throughout the text. Furthermore it contains a number of subtle points. Thus, we pause to make a few additional observations.

The σ-algebra $\mathcal{F}$ is the collection of all *events* or *measurable sets*. These are the subsets $A \subseteq \Omega$ for which $\mathbf{P}(A)$ is well-defined. We know from Proposition 1.2.6 that in general $\mathcal{F}$ might not contain *all* subsets of Ω, though we still expect it to contain most of the subsets that come up naturally.

To say that $\mathcal{F}$ is closed under the formation of complements and countable unions and countable intersections means, more precisely, that
(i) For any subset $A \subseteq \Omega$, if $A \in \mathcal{F}$, then $A^C \in \mathcal{F}$;
(ii) For any countable (or finite) collection of subsets $A_1, A_2, A_3, \ldots \subseteq \Omega$, if $A_i \in \mathcal{F}$ for each i, then the union $A_1 \cup A_2 \cup A_3 \cup \ldots \in \mathcal{F}$;
(iii) For any countable (or finite) collection of subsets $A_1, A_2, A_3, \ldots \subseteq \Omega$, if $A_i \in \mathcal{F}$ for each i, then the intersection $A_1 \cap A_2 \cap A_3 \cap \ldots \in \mathcal{F}$.
Like for countable additivity, the reason we require $\mathcal{F}$ to be closed under countable operations is to allow for taking limits, etc., when studying probability theory. Also like for additivity, we cannot extend the definition to

---

[*]For the definitions of complements, unions, intersections, etc., see Subsection A.1 on page 199.

require that $\mathcal{F}$ be closed under *uncountable* unions; in this case, for the example of Subsection 1.2 above, $\mathcal{F}$ would contain *every* subset A, since every subset can be written as $A = \bigcup_{x \in A}\{x\}$ and since the singleton sets $\{x\}$ are all in $\mathcal{F}$.

There is some redundancy in the definition above. For example, it follows from de Morgan's Laws (see Subsection A.1) that if $\mathcal{F}$ is closed under complement and countable unions, then it is *automatically* closed under countable intersections. Similarly, it follows from countable additivity that we must have $\mathbf{P}(\emptyset) = 0$, and that (once we know that $\mathbf{P}(\Omega) = 1$ and $\mathbf{P}(A) \geq 0$ for all $A \in \mathcal{F}$) we must have $\mathbf{P}(A) \leq 1$.

More generally, from additivity we have $\mathbf{P}(A) + \mathbf{P}(A^C) = \mathbf{P}(A \cup A^C) = \mathbf{P}(\Omega) = 1$, whence

$$\mathbf{P}(A^C) = 1 - \mathbf{P}(A), \tag{2.1.1}$$

a fact that will be used often. Similarly, if $A \subseteq B$, then since $B = A \mathbin{\dot{\cup}} (B \backslash A)$ (where $\dot{\cup}$ means disjoint union), we have that $\mathbf{P}(B) = \mathbf{P}(A) + \mathbf{P}(B \backslash A) \geq \mathbf{P}(A)$, i.e.

$$\mathbf{P}(A) \leq \mathbf{P}(B) \qquad \text{whenever } A \subseteq B, \tag{2.1.2}$$

which is the *monotonicity* property of probability measures.

Also, if $A, B \in \mathcal{F}$, then

$$
\begin{aligned}
\mathbf{P}(A \cup B) &= \mathbf{P}\big[(A \backslash B) \mathbin{\dot{\cup}} (B \backslash A) \mathbin{\dot{\cup}} (A \cap B)\big] \\
&= \mathbf{P}(A \backslash B) + \mathbf{P}(B \backslash A) + \mathbf{P}(A \cap B) \\
&= \mathbf{P}(A) - \mathbf{P}(A \cap B) + \mathbf{P}(B) - \mathbf{P}(A \cap B) + \mathbf{P}(A \cap B) \\
&= \mathbf{P}(A) + \mathbf{P}(B) - \mathbf{P}(A \cap B),
\end{aligned}
$$

the *principle of inclusion-exclusion*. For a generalisation see Exercise 4.5.7.

Finally, for any sequence $A_1, A_2, \ldots \in \mathcal{F}$ (whether disjoint or not), we have by countable additivity and monotonicity that

$$
\begin{aligned}
\mathbf{P}(A_1 \cup A_2 \cup A_3 \cup \ldots) &= \mathbf{P}\big[A_1 \mathbin{\dot{\cup}} (A_2 \backslash A_1) \mathbin{\dot{\cup}} (A_3 \backslash A_2 \backslash A_1) \mathbin{\dot{\cup}} \ldots\big] \\
&= \mathbf{P}(A_1) + \mathbf{P}(A_2 \backslash A_1) + \mathbf{P}(A_3 \backslash A_2 \backslash A_1) + \ldots \\
&\leq \mathbf{P}(A_1) + \mathbf{P}(A_2) + \mathbf{P}(A_3) + \ldots
\end{aligned}
$$

which is the *countable subadditivity* property of probability measures.

2.2. Constructing probability triples.

We clarify the definition of Subsection 2.1 with a simple example. Let us again consider the **Poisson**(5) distribution considered in Subsection 1.1. In this case, the sample space Ω would consist of all the non-negative integers:

$\Omega = \{0, 1, 2, \ldots\}$. Also, the σ-algebra $\mathcal{F}$ would consist of *all* subsets of Ω. Finally, the probability measure $\mathbf{P}$ would be defined, for any $A \in \mathcal{F}$, by

$$\mathbf{P}(A) = \sum_{k \in A} e^{-5} 5^k / k! \,.$$

It is straightforward to check that $\mathcal{F}$ is indeed a σ-algebra (it contains *all* subsets of Ω, so it's closed under *any* set operations), and that $\mathbf{P}$ is a probability measure defined on $\mathcal{F}$ (the additivity following since if A and B are disjoint, then $\sum_{k \in A \cup B}$ is the same as $\sum_{k \in A} + \sum_{k \in B}$).

So in the case of **Poisson**(5), we see that it is entirely straightforward to construct an appropriate probability triple. The construction is similarly straightforward for any *discrete* probability space, i.e. any space for which the sample space Ω is finite or countable. We record this as follows.

Theorem 2.2.1.   *Let Ω be a finite or countable non-empty set. Let $p : \Omega \to [0, 1]$ be any function satisfying $\sum_{\omega \in \Omega} p(\omega) = 1$. Then there is a valid probability triple $(\Omega, \mathcal{F}, \mathbf{P})$ where $\mathcal{F}$ is the collection of all subsets of Ω, and for $A \in \mathcal{F}$, $\mathbf{P}(A) = \sum_{\omega \in A} p(\omega)$.*

Example 2.2.2.   Let Ω be any finite non-empty set, $\mathcal{F}$ be the collection of all subsets of Ω, and $\mathbf{P}(A) = |A| / |\Omega|$ for all $A \in \mathcal{F}$ (where $|A|$ is the cardinality of the set A). Then $(\Omega, \mathcal{F}, \mathbf{P})$ is a valid probability triple, called the *uniform distribution on Ω*, written **Uniform**(Ω).

However, if the sample space is *not* countable, then the situation is considerably more complex, as seen in Subsection 1.2. How can we formally define a probability triple $(\Omega, \mathcal{F}, \mathbf{P})$ which corresponds to, say, the **Uniform**$[0, 1]$ distribution?

It seems clear that we should choose $\Omega = [0, 1]$. But what about $\mathcal{F}$? We know from Proposition 1.2.6 that $\mathcal{F}$ cannot contain *all* subsets of Ω, but it should certainly contain all the intervals $[a, b]$, $[a, b)$, etc. That is, we must have $\mathcal{F} \supseteq \mathcal{J}$, where

$$\mathcal{J} = \{\text{all intervals contained in } [0, 1]\}$$

and where "intervals" is understood to include all the open / closed / half-open / singleton / empty intervals.

Exercise 2.2.3.   Prove that the above collection $\mathcal{J}$ is a *semialgebra* of subsets of Ω, meaning that it contains $\emptyset$ and Ω, it is closed under finite intersection, and the complement of any element of $\mathcal{J}$ is equal to a finite disjoint union of elements of $\mathcal{J}$.

Since $\mathcal{J}$ is only a semialgebra, how can we create a σ-algebra? As a first try, we might consider

$$\mathcal{B}_0 \; = \; \{\text{all finite unions of elements of } \mathcal{J}\}. \tag{2.2.4}$$

(After all, by additivity we already know how to define $\mathbf{P}$ on $\mathcal{B}_0$.) However, $\mathcal{B}_0$ is not a σ-algebra:

Exercise 2.2.5.   **(a)** Prove that $\mathcal{B}_0$ is an *algebra* (or, *field*) of subsets of Ω, meaning that it contains Ω and $\emptyset$, and is closed under the formation of complements and of *finite* unions and intersections.
(b) Prove that $\mathcal{B}_0$ is <u>not</u> a σ-algebra.

As a second try, we might consider

$$\mathcal{B}_1 \; = \; \{\text{all finite or countable unions of elements of } \mathcal{J}\}. \tag{2.2.6}$$

Unfortunately, $\mathcal{B}_1$ is still not a σ-algebra (Exercise 2.4.7).

Thus, the construction of $\mathcal{F}$, and of $\mathbf{P}$, presents serious challenges. To deal with them, we next prove a very general theorem about constructing probability triples.

2.3. The Extension Theorem.

The following theorem is of fundamental importance in constructing complicated probability triples. Recall the definition of *semialgebra* from Exercise 2.2.3.

Theorem 2.3.1.   *(The Extension Theorem.) Let $\mathcal{J}$ be a semialgebra of subsets of Ω. Let $\mathbf{P} : \mathcal{J} \to [0, 1]$ with $\mathbf{P}(\emptyset) = 0$ and $\mathbf{P}(\Omega) = 1$, satisfying the finite superadditivity property that*

$$\mathbf{P}\left(\bigcup_{i=1}^{k} A_i\right) \geq \sum_{i=1}^{k} \mathbf{P}(A_i) \quad \text{whenever } A_1, \ldots, A_k \in \mathcal{J}, \text{ and } \bigcup_{i=1}^{k} A_i \in \mathcal{J},$$
$$\text{and the } \{A_i\} \text{ are disjoint}, \tag{2.3.2}$$

and also the countable monotonicity property that

$$\mathbf{P}(A) \; \leq \; \sum_{n} \mathbf{P}(A_n) \quad \text{for } A, A_1, A_2, \ldots \in \mathcal{J} \text{ with } A \subseteq \bigcup_{n} A_n. \tag{2.3.3}$$

Then there is a σ-algebra $\mathcal{M} \supseteq \mathcal{J}$, and a countably additive probability measure $\mathbf{P}^*$ on $\mathcal{M}$, such that $\mathbf{P}^*(A) = \mathbf{P}(A)$ for all $A \in \mathcal{J}$. (That is,

$(\Omega, \mathcal{M}, \mathbf{P}^*)$ *is a valid probability triple, which agrees with our previous probabilities on* $\mathcal{J}$.*)*

Remark. Of course, the conclusions of Theorem 2.3.1 imply that (2.3.2) must actually hold with *equality*. However, (2.3.2) need only be verified as an *inequality* to apply Theorem 2.3.1.

Theorem 2.3.1 provides precisely what we need: a way to construct complicated probability triples on a full σ-algebra, using only probabilities defined on the much simpler subsets (e.g., intervals) in $\mathcal{J}$.

However, it is not clear how to even *start* proving this theorem. Indeed, how could we begin to define $\mathbf{P}(A)$ for *all* A in a σ-algebra? The key is given by *outer measure* $\mathbf{P}^*$, defined by

$$\mathbf{P}^*(A) = \inf_{\substack{A_1, A_2, \ldots \in \mathcal{J} \\ A \subseteq \bigcup_i A_i}} \sum_i \mathbf{P}(A_i), \qquad A \subseteq \Omega. \qquad (2.3.4)$$

That is, we define $\mathbf{P}^*(A)$, for *any* subset $A \subseteq \Omega$, to be the infimum of sums of $\mathbf{P}(A_i)$, where $\{A_i\}$ is any countable collection of elements of the original semialgebra $\mathcal{J}$ whose union contains A. In other words, we use the values of $\mathbf{P}(A)$ for $A \in \mathcal{J}$, to help us define $\mathbf{P}^*(A)$ for any $A \subseteq \Omega$. Of course, we know that $\mathbf{P}^*$ will *not* necessarily be a proper probability measure for all $A \subseteq \Omega$; for example, this is not possible for **Uniform**$[0, 1]$ by Proposition 1.2.6. However, it is still useful that $\mathbf{P}^*(A)$ is at least *defined* for all $A \subseteq \Omega$. We shall eventually show that $\mathbf{P}^*$ is indeed a probability measure on some σ-algebra $\mathcal{M}$, and that $\mathbf{P}^*$ is an extension of $\mathbf{P}$.

To continue, we note a few simple properties of $\mathbf{P}^*$. Firstly, we clearly have $\mathbf{P}^*(\emptyset) = 0$; indeed, we can simply take $A_i = \emptyset$ for each i in the definition (2.3.4). Secondly, $\mathbf{P}^*$ is clearly *monotone*; indeed, if $A \subseteq B$ then the infimum (2.3.4) for $\mathbf{P}^*(A)$ includes all choices of $\{A_i\}$ which work for $\mathbf{P}^*(B)$ plus many more besides, so that $\mathbf{P}^*(A) \leq \mathbf{P}^*(B)$. We also have:

Lemma 2.3.5. $\mathbf{P}^*$ *is an extension of* $\mathbf{P}$, *i.e.* $\mathbf{P}^*(A) = \mathbf{P}(A)$ *for all* $A \in \mathcal{J}$.

Proof. Let $A \in \mathcal{J}$. It follows from (2.3.3) that $\mathbf{P}^*(A) \geq \mathbf{P}(A)$. On the other hand, choosing $A_1 = A$ and $A_i = \emptyset$ for $i > 1$ in the definition (2.3.4) shows by (A.4.1) that $\mathbf{P}^*(A) \leq \mathbf{P}(A)$. ∎

Lemma 2.3.6. $\mathbf{P}^*$ *is countably subadditive, i.e.*

$$\mathbf{P}^*\left(\bigcup_{n=1}^{\infty} B_n\right) \leq \sum_{n=1}^{\infty} \mathbf{P}^*(B_n) \quad \text{for any} \quad B_1, B_2, \ldots \subseteq \Omega.$$

Proof.   Let $B_1, B_2, \ldots \subseteq \Omega$. From the definition (2.3.4), we see that for any $\epsilon > 0$, we can find (cf. Proposition A.4.2) a collection $\{C_{nk}\}_{k=1}^{\infty}$ for each $n \in \mathbf{N}$, with $C_{nk} \in \mathcal{J}$, such that $B_n \subseteq \bigcup_k C_{nk}$ and $\sum_k \mathbf{P}(C_{nk}) \leq \mathbf{P}^*(B_n) + \epsilon 2^{-n}$. But then the overall collection $\{C_{nk}\}_{n,k=1}^{\infty}$ contains $\bigcup_{n=1}^{\infty} B_n$. It follows that $\mathbf{P}^*\left(\bigcup_{n=1}^{\infty} B_n\right) \leq \sum_{n,k} \mathbf{P}(C_{nk}) \leq \sum_n \mathbf{P}^*(B_n) + \epsilon$. Since this is true for any $\epsilon > 0$, we must have (cf. Proposition A.3.1) that $\mathbf{P}^*\left(\bigcup_{n=1}^{\infty} B_n\right) \leq \sum_n \mathbf{P}^*(B_n)$, as claimed. ∎

We now set

$$\mathcal{M} = \{A \subseteq \Omega; \ \mathbf{P}^*(A \cap E) + \mathbf{P}^*(A^C \cap E) = \mathbf{P}^*(E) \ \forall E \subseteq \Omega\}. \quad (2.3.7)$$

That is, $\mathcal{M}$ is the set of all subsets A with the property that $\mathbf{P}^*$ is additive on the union of $A \cap E$ with $A^C \cap E$, for *all* subsets E. Note that by subadditivity we always have $\mathbf{P}^*(A \cap E) + \mathbf{P}^*(A^C \cap E) \geq \mathbf{P}^*(E)$, so (2.3.7) is equivalent to

$$\mathcal{M} = \{A \subseteq \Omega; \ \mathbf{P}^*(A \cap E) + \mathbf{P}^*(A^C \cap E) \leq \mathbf{P}^*(E) \ \forall E \subseteq \Omega\}, \quad (2.3.8)$$

which is sometimes helpful. Furthermore, $\mathbf{P}^*$ is countably additive on $\mathcal{M}$:

Lemma 2.3.9.   If $A_1, A_2, \ldots \in \mathcal{M}$ are disjoint, then $\mathbf{P}^*\left(\bigcup_n A_n\right) = \sum_n \mathbf{P}^*(A_n)$.

Proof.   If A_1 and A_2 are disjoint, with $A_1 \in \mathcal{M}$, then

$$\begin{aligned}
&\mathbf{P}^*(A_1 \cup A_2) \\
&\quad = \mathbf{P}^*\left(A_1 \cap (A_1 \cup A_2)\right) + \mathbf{P}^*\left(A_1^C \cap (A_1 \cup A_2)\right) \qquad \text{since } A_1 \in \mathcal{M} \\
&\quad = \mathbf{P}^*(A_1) + \mathbf{P}^*(A_2) \qquad \text{since } A_1, A_2 \text{ disjoint}.
\end{aligned}$$

Hence, by induction, the lemma holds for any *finite* collection of A_i.

Then, with countably many disjoint $A_i \in \mathcal{M}$, we see that for any $m \in \mathbf{N}$,

$$\sum_{n \leq m} \mathbf{P}^*(A_n) = \mathbf{P}^*\left(\bigcup_{n \leq m} A_n\right) \leq \mathbf{P}^*\left(\bigcup_n A_n\right),$$

where the inequality follows from monotonicity. Since this is true for any $m \in \mathbf{N}$, we have (cf. Proposition A.3.6) that $\sum_n \mathbf{P}^*(A_n) \leq \mathbf{P}^*\left(\bigcup_n A_n\right)$. On the other hand, by subadditivity we have $\sum_n \mathbf{P}^*(A_n) \geq \mathbf{P}^*\left(\bigcup_n A_n\right)$. Hence, the lemma holds for countably many A_i as well. ∎

The plan now is to show that $\mathcal{M}$ is a σ-algebra which contains $\mathcal{J}$. We break up the proof into a number of lemmas.

Lemma 2.3.10. $\mathcal{M}$ is an algebra, i.e. $\Omega \in \mathcal{M}$ and $\mathcal{M}$ is closed under complement and finite intersection (and hence also finite union).

Proof. It is immediate from (2.3.7) that $\mathcal{M}$ contains Ω and is closed under complement. For the statement about finite intersections, suppose $A, B \in \mathcal{M}$. Then, for any $E \subseteq \Omega$, using subadditivity,

$$
\begin{aligned}
&\mathbf{P}^*\left((A \cap B) \cap E\right) + \mathbf{P}^*\left((A \cap B)^C \cap E\right) \\
&= \mathbf{P}^*\left(A \cap B \cap E\right) \\
&\qquad\qquad + \mathbf{P}^*\left((A^C \cap B \cap E) \cup (A \cap B^C \cap E) \cup (A^C \cap B^C \cap E)\right) \\
&\leq \mathbf{P}^*\left(A \cap B \cap E\right) + \mathbf{P}^*\left(A^C \cap B \cap E\right) \\
&\qquad\qquad + \mathbf{P}^*\left(A \cap B^C \cap E\right) + \mathbf{P}^*\left(A^C \cap B^C \cap E\right)) \\
&= \mathbf{P}^*(B \cap E) + \mathbf{P}^*(B^C \cap E) \qquad \text{since } A \in \mathcal{M} \\
&= \mathbf{P}^*(E) \qquad \text{since } B \in \mathcal{M}.
\end{aligned}
$$

Hence, by (2.3.8), $A \cap B \in \mathcal{M}$. ∎

Lemma 2.3.11. Let $A_1, A_2, \ldots \in \mathcal{M}$ be disjoint. For each $m \in \mathbf{N}$, let $B_m = \bigcup_{n \leq m} A_n$. Then for all $m \in \mathbf{N}$, and for all $E \subseteq \Omega$, we have

$$
\mathbf{P}^*(E \cap B_m) = \sum_{n \leq m} \mathbf{P}^*(E \cap A_n). \tag{2.3.12}
$$

Proof. We use induction on m. Indeed, the statement is trivially true when $m = 1$. Assuming it true for some particular value of m, and noting that $B_m \cap B_{m+1} = B_m$ and $B_m^C \cap B_{m+1} = A_{m+1}$, and that $B_m \in \mathcal{M}$ by Proposition 2.3.10, we have that

$$
\begin{aligned}
&\mathbf{P}^*(E \cap B_{m+1}) \\
&= \mathbf{P}^*(B_m \cap E \cap B_{m+1}) + \mathbf{P}^*(B_m^C \cap E \cap B_{m+1}) \qquad \text{since } B_m \in \mathcal{M} \\
&= \mathbf{P}^*(E \cap B_m) + \mathbf{P}^*(E \cap A_{m+1}) \\
&= \sum_{n \leq m+1} \mathbf{P}^*(E \cap A_n) \qquad \text{by the induction hypothesis},
\end{aligned}
$$

thus completing the induction proof. ∎

Lemma 2.3.13. Let $A_1, A_2, \ldots \in \mathcal{M}$ be disjoint. Then $\bigcup_n A_n \in \mathcal{M}$.

Proof. For each $m \in \mathbf{N}$, let $B_m = \bigcup_{n \leq m} A_n$. Then for any $m \in \mathbf{N}$ and any $E \subseteq \Omega$,

$$
\begin{aligned}
\mathbf{P}^*(E) &= \mathbf{P}^*(E \cap B_m) + \mathbf{P}^*(E \cap B_m^C) \qquad \text{since } B_m \in \mathcal{M} \\
&= \sum_{n \leq m} \mathbf{P}^*(E \cap A_n) + \mathbf{P}^*(E \cap B_m^C) \qquad \text{by (2.3.12)} \\
&\geq \sum_{n \leq m} \mathbf{P}^*(E \cap A_n) + \mathbf{P}^*(E \cap (\bigcup A_n)^C),
\end{aligned}
$$

where the inequality follows by monotonicity since $(\bigcup A_n)^C \subseteq B_m^C$. This is true for any $m \in \mathbf{N}$, so it implies (cf. Proposition A.3.6) that

$$\mathbf{P}^*(E) \geq \sum_n \mathbf{P}^*(E \cap A_n) + \mathbf{P}^* \left(E \cap \left(\bigcup_n A_n \right)^C \right)$$

$$\geq \mathbf{P}^* \left(E \cap \left(\bigcup_n A_n \right) \right) + \mathbf{P}^* \left(E \cap \left(\bigcup_n A_n \right)^C \right),$$

where the final inequality follows by subadditivity. Hence, by (2.3.8) we have $\bigcup_n A_n \in \mathcal{M}$. ∎

Lemma 2.3.14. $\mathcal{M}$ is a σ-algebra.

Proof.   In light of Lemma 2.3.10, it suffices to show that $\bigcup_n A_n \in \mathcal{M}$ whenever $A_1, A_2, \ldots \in \mathcal{M}$. Let $D_1 = A_1$, and $D_i = A_i \cap A_1^C \cap \ldots \cap A_{i-1}^C$ for $i \geq 2$. Then $\{D_i\}$ are disjoint, with $\bigcup_i D_i = \bigcup_i A_i$, and with $D_i \in \mathcal{M}$ by Lemma 2.3.10. Hence, by Lemma 2.3.13, $\bigcup_i D_i \in \mathcal{M}$, i.e. $\bigcup_i A_i \in \mathcal{M}$. ∎

Lemma 2.3.15. $\mathcal{J} \subseteq \mathcal{M}$.

Proof.   Let $A \in \mathcal{J}$. Then since $\mathcal{J}$ is a semialgebra, we can write $A^C = J_1 \mathbin{\dot{\cup}} \ldots \mathbin{\dot{\cup}} J_k$ for some disjoint $J_1, \ldots, J_k \in \mathcal{J}$. Also, for any $E \subseteq \Omega$ and $\epsilon > 0$, by the definition (2.3.4) we can find (cf. Proposition A.4.2) $A_1, A_2, \ldots \in \mathcal{J}$ with $E \subseteq \bigcup_n A_n$ and $\sum_n \mathbf{P}(A_n) \leq \mathbf{P}^*(E) + \epsilon$. Then

$$
\begin{aligned}
&\mathbf{P}^*(E \cap A) + \mathbf{P}^*(E \cap A^C) \\
&\leq \mathbf{P}^*\left(\left(\textstyle\bigcup_n A_n \right) \cap A \right) + \mathbf{P}^*\left(\left(\textstyle\bigcup_n A_n \right) \cap A^C \right) && \text{by monotonicity} \\
&= \mathbf{P}^*\left(\textstyle\bigcup_n (A_n \cap A) \right) + \mathbf{P}^*\left(\textstyle\bigcup_n \bigcup_{i=1}^k (A_n \cap J_i) \right) \\
&\leq \textstyle\sum_n \mathbf{P}^*(A_n \cap A) + \sum_n \sum_{i=1}^k \mathbf{P}^*(A_n \cap J_i) && \text{by subadditivity} \\
&= \textstyle\sum_n \mathbf{P}(A_n \cap A) + \sum_n \sum_{i=1}^k \mathbf{P}(A_n \cap J_i) && \text{since } \mathbf{P}^* = \mathbf{P} \text{ on } \mathcal{J} \\
&= \textstyle\sum_n \left(\mathbf{P}(A_n \cap A) + \sum_{i=1}^k \mathbf{P}(A_n \cap J_i) \right) \\
&\leq \textstyle\sum_n \mathbf{P}(A_n) && \text{by (2.3.2)} \\
&\leq \mathbf{P}^*(E) + \epsilon && \text{by assumption}.
\end{aligned}
$$

This is true for any $\epsilon > 0$, hence (cf. Proposition A.3.1) we have $\mathbf{P}^*(E \cap A) + \mathbf{P}^*(E \cap A^C) \leq \mathbf{P}^*(E)$, for any $E \subseteq \Omega$. Hence, from (2.3.8), we have $A \in \mathcal{M}$. This holds for any $A \in \mathcal{J}$, hence $\mathcal{J} \subseteq \mathcal{M}$. ∎

With all those lemmas behind us, we are now, finally, able to complete the proof of the Extension Theorem.

Proof of Theorem 2.3.1.   Lemmas 2.3.5, 2.3.9, 2.3.14, and 2.3.15 together show that $\mathcal{M}$ is a σ-algebra containing $\mathcal{J}$, that $\mathbf{P}^*$ is a probability measure on $\mathcal{M}$, and that $\mathbf{P}^*$ is an extension of $\mathbf{P}$.   ∎

Exercise 2.3.16.   Prove that the extension $(\Omega, \mathcal{M}, \mathbf{P}^*)$ constructed in the proof of Theorem 2.3.1 must be *complete*, meaning that if $A \in \mathcal{M}$ with $\mathbf{P}^*(A) = 0$, and if $B \subseteq A$, then $B \in \mathcal{M}$. (It then follows from monotonicity that $\mathbf{P}^*(B) = 0$.)

2.4. Constructing the Uniform$[0,1]$ distribution.

Theorem 2.3.1 allows us to automatically construct valid probability triples which take particular values on particular sets. We now use this to construct the **Uniform**$[0,1]$ distribution. We begin by letting $\Omega = [0,1]$, and again setting

$$\mathcal{J} = \{\text{all intervals contained in } [0,1]\}, \tag{2.4.1}$$

where again "intervals" is understood to include all the open, closed, half-open, and singleton intervals contained in $[0,1]$, and also the empty set $\emptyset$. Then $\mathcal{J}$ is a semialgebra by Exercise 2.2.3.

For $I \in \mathcal{J}$, we let $\mathbf{P}(I)$ be the length of I. Thus $\mathbf{P}(\emptyset) = 0$ and $\mathbf{P}(\Omega) = 1$. We now proceed to verify (2.3.2) and (2.3.3).

Proposition 2.4.2.   *The above definition of $\mathcal{J}$ and $\mathbf{P}$ satisfies (2.3.2), with equality.*

Proof.   Let $I_1, \ldots, I_k$ be disjoint intervals contained in $[0,1]$, whose union is some interval I_0. For $0 \le j \le k$, write a_j for the left end-point of I_j, and b_j for the right end-point of I_j. The assumptions imply that by re-ordering, we can ensure that $a_0 = a_1 \le b_1 = a_2 \le b_2 = a_3 \le \ldots \le b_k = b_0$. Then

$$\sum_j \mathbf{P}(I_j) = \sum_j (b_j - a_j) = b_k - a_1 = b_0 - a_0 = \mathbf{P}(I_0).\qquad ∎$$

The verification of (2.3.3) for this $\mathcal{J}$ and $\mathbf{P}$ is a bit more involved:

Exercise 2.4.3.   (a) Prove that if $I_1, I_2, \ldots, I_n$ is a finite collection of intervals, and if $\bigcup_{j=1}^n I_j \supseteq I$ for some interval I, then $\sum_{j=1}^n \mathbf{P}(I_j) \ge \mathbf{P}(I)$. [Hint: Imitate the proof of Proposition 2.4.2.]
(b) Prove that if $I_1, I_2, \ldots$ is a countable collection of *open* intervals, and if $\bigcup_{j=1}^\infty I_j \supseteq I$ for some closed interval I, then $\sum_{j=1}^\infty \mathbf{P}(I_j) \ge \mathbf{P}(I)$. [Hint:

You may use the *Heine-Borel Theorem*, which says that if a collection of open intervals contain a closed interval, then some finite sub-collection of the open intervals also contains the closed interval.]

(c) Verify (2.3.3), i.e. prove that if $I_1, I_2, \ldots$ is *any* countable collection of intervals, and if $\bigcup_{j=1}^{\infty} I_j \supseteq I$ for *any* interval I, then $\sum_{j=1}^{\infty} \mathbf{P}(I_j) \geq \mathbf{P}(I)$. [Hint: Extend the interval I_j by $\epsilon \, 2^{-j}$ at each end, and decrease I by ϵ at each end, while making I_j open and I closed. Then use part (b).]

In light of Proposition 2.4.2 and Exercise 2.4.3, we can apply Theorem 2.3.1 to conclude the following:

Theorem 2.4.4.   *There exists a probability triple* $(\Omega, \mathcal{M}, \mathbf{P}^*)$ *such that* $\Omega = [0, 1]$, $\mathcal{M}$ *contains all intervals in* $[0, 1]$, *and for any interval* $I \subseteq [0, 1]$, $\mathbf{P}^*(I)$ *is the length of* I.

This probability triple is called either the uniform distribution on $[0, 1]$, or *Lebesgue measure* on $[0, 1]$. Depending on the context, we sometimes write the probability measure $\mathbf{P}^*$ as $\mathbf{P}$ or as λ.

Remark.   Let $\mathcal{B} = \sigma(\mathcal{J})$ be the σ-algebra *generated* by $\mathcal{J}$, i.e. the smallest σ-algebra containing $\mathcal{J}$. (The collection $\mathcal{B}$ is called the *Borel σ-algebra* of subsets of $[0, 1]$, and the elements of $\mathcal{B}$ are called *Borel sets*.) Clearly, we must have $\mathcal{M} \supseteq \mathcal{B}$. In this case, it can be shown that $\mathcal{M}$ is in fact much bigger than $\mathcal{B}$; it even has larger cardinality. Furthermore, it turns out that Lebesgue measure restricted to $\mathcal{B}$ is not complete, though on $\mathcal{M}$ it is (by Exercise 2.3.16). In addition to the Borel subsets of $[0, 1]$, we shall also have occasion to refer to the *Borel σ-algebra of subsets of* $\mathbf{R}$, defined to be the smallest σ-algebra of subsets of $\mathbf{R}$ which includes all intervals.

Exercise 2.4.5.   Let $\mathcal{A} = \big\{ (-\infty, x]; \ x \in \mathbf{R} \big\}$. Prove that $\sigma(\mathcal{A}) = \mathcal{B}$, i.e. that the smallest σ-algebra of subsets of $\mathbf{R}$ which contains $\mathcal{A}$ is equal to the Borel σ-algebra of subsets of $\mathbf{R}$. [Hint: Does $\sigma(\mathcal{A})$ include all intervals?]

Writing λ for Lebesgue measure on $[0, 1]$, we know that $\lambda\{x\} = 0$ for any singleton set $\{x\}$. It follows by countable additivity that $\lambda(A) = 0$ for any set A which is *countable*. This includes (cf. Subsection A.2) the rational numbers, the integer roots of the rational numbers, the algebraic numbers, etc. That is, if X is uniformly distributed on $[0, 1]$, then $\mathbf{P}(X$ is rational$) = 0$, and $\mathbf{P}(X^n$ is rational for some $n \in \mathbf{N}) = 0$, and $\mathbf{P}(X$ is algebraic$) = 0$, and so on.

There also exist uncountable sets which have Lebesgue measure 0. The simplest example is the *Cantor set* K, defined as follows (see Figure 2.4.6). We begin with the interval $[0, 1]$. We then *remove* the open interval con-

Figure 2.4.6. Constructing the Cantor set K.

sisting of the middle third $(1/3, 2/3)$. We then *remove* the open middle thirds of each of the two pieces, i.e. we remove $(1/9, 2/9)$ and $(7/9, 8/9)$. We then remove the four open middle thirds $(1/27, 2/27)$, $(7/27, 8/27)$, $(19/27, 20/27)$, and $(25/27, 26/27)$ of the remaining pieces. We continue inductively, at the nth stage removing the 2^{n-1} middle thirds of all remaining sub-intervals, each of length $1/3^n$. The Cantor set K is defined to be everything that is left over, after we have removed all these middle thirds.

Now, the *complement* of the Cantor set has Lebesgue measure given by $\lambda(K^C) = 1/3 + 2(1/9) + 4(1/27) + \ldots = \sum_{n=1}^{\infty} 2^{n-1}/3^n = 1$. Hence, by (2.1.1), $\lambda(K) = 1 - 1 = 0$.

On the other hand, K is uncountable. Indeed, for each point $x \in K$, let $d_n(x) = 0$ or 1 depending on whether, at the nth stage of the construction of K, x was to the left or the right of the nearest open interval removed. Then define the function $f : K \to [0, 1]$ by $f(x) = \sum_{n=1}^{\infty} d_n(x)\, 2^{-n}$. It is easily checked that $f(K) = [0, 1]$, i.e. that f maps K *onto* $[0, 1]$. Since $[0, 1]$ is uncountable, this means that K must also be uncountable.

Remark. The Cantor set is also equal to the set of all numbers in $[0, 1]$ which have a *base-3 expansion* that does not contain the digit 1. That is, $K = \left\{ \sum_{n=1}^{\infty} c_n 3^{-n} : \text{each } c_n \in \{0, 2\} \right\}$.

Exercise 2.4.7.   (a) Prove that $K, K^C \in \mathcal{B}$, where $\mathcal{B}$ are the Borel subsets of $[0, 1]$.
(b) Prove that $K, K^C \in \mathcal{M}$, where $\mathcal{M}$ is the σ-algebra of Theorem 2.4.4.
(c) Prove that $K^C \in \mathcal{B}_1$, where $\mathcal{B}_1$ is defined by (2.2.6).
(d) Prove that $K \notin \mathcal{B}_1$.
(e) Prove that $\mathcal{B}_1$ is not a σ-algebra.

On the other hand, from Proposition 1.2.6 we know that:

Proposition 2.4.8.   *For the probability triple* $(\Omega, \mathcal{M}, \mathbf{P}^*)$ *of Theorem 2.4.4 corresponding to Lebesgue measure on* $[0, 1]$, *there exists at least one subset* $H \subseteq \Omega$ *with* $H \notin \mathcal{M}$.

2.5. Extensions of the Extension Theorem.

The Extension Theorem (Theorem 2.3.1) will be our main tool for proving the existence of complicated probability triples. While (2.3.2) is generally easy to verify, (2.3.3) can be more challenging. Thus, we present some alternative formulations here.

Corollary 2.5.1.   *Let* $\mathcal{J}$ *be a semialgebra of subsets of* Ω. *Let* $\mathbf{P} : \mathcal{J} \to [0, 1]$ *with* $\mathbf{P}(\emptyset) = 0$ *and* $\mathbf{P}(\Omega) = 1$, *satisfying (2.3.2), and the "monotonicity on* $\mathcal{J}$*" property that*

$$\mathbf{P}(A) \le \mathbf{P}(B) \quad \text{whenever} \quad A, B \in \mathcal{J} \text{ with } A \subseteq B, \tag{2.5.2}$$

and also the "countable subadditivity on $\mathcal{J}$*" property that*

$$\mathbf{P}\left(\bigcup_n B_n\right) \le \sum_n \mathbf{P}(B_n) \quad \text{for } B_1, B_2, \ldots \in \mathcal{J} \text{ with } \bigcup_n B_n \in \mathcal{J}. \tag{2.5.3}$$

Then there is a σ-*algebra* $\mathcal{M} \supseteq \mathcal{J}$, *and a countably additive probability measure* $\mathbf{P}^*$ *on* $\mathcal{M}$, *such that* $\mathbf{P}^*(A) = \mathbf{P}(A)$ *for all* $A \in \mathcal{J}$.

Proof.   In light of Theorem 2.3.1, we need only verify (2.3.3). To that end, let $A, A_1, A_2, \ldots \in \mathcal{J}$ with $A \subseteq \bigcup_n A_n$. Set $B_n = A \cap A_n$. Then since $A \subseteq \bigcup_n A_n$, we have $A = \bigcup_n (A \cap A_n) = \bigcup_n B_n$, whence (2.5.3) and (2.5.2) give that

$$\mathbf{P}(A) = \mathbf{P}\left(\bigcup_n B_n\right) \le \sum_n \mathbf{P}(B_n) \le \sum_n \mathbf{P}(A_n). \qquad \blacksquare$$

Another version assumes countable additivity of $\mathbf{P}$ on $\mathcal{J}$:

Corollary 2.5.4.   *Let* $\mathcal{J}$ *be a semialgebra of subsets of* Ω. *Let* $\mathbf{P} : \mathcal{J} \to [0, 1]$ *with* $\mathbf{P}(\Omega) = 1$, *satisfying the countable additivity property that*

$$\mathbf{P}\left(\bigcup_n D_n\right) = \sum_n \mathbf{P}(D_n) \quad \text{for } D_1, D_2, \ldots \in \mathcal{J} \text{ disjoint with } \bigcup_n D_n \in \mathcal{J}. \tag{2.5.5}$$

Then there is a σ-*algebra* $\mathcal{M} \supseteq \mathcal{J}$, *and a countably additive probability measure* $\mathbf{P}^*$ *on* $\mathcal{M}$, *such that* $\mathbf{P}^*(A) = \mathbf{P}(A)$ *for all* $A \in \mathcal{J}$.

Proof. Note that (2.5.5) immediately implies (2.3.2) (with equality), and that $\mathbf{P}(\emptyset) = 0$. Hence, in light of Corollary 2.5.1, we need only verify (2.5.2) and (2.5.3).

For (2.5.2), let $A, B \in \mathcal{J}$ with $A \subseteq B$. Since $\mathcal{J}$ is a semialgebra, we can write $A^C = J_1 \dot{\cup} \ldots \dot{\cup} J_k$, for some disjoint $J_1, \ldots, J_k \in \mathcal{J}$. Then using (2.5.5),

$$\mathbf{P}(B) = \mathbf{P}(A) + \mathbf{P}(B \cap J_1) + \ldots + \mathbf{P}(B \cap J_k) \geq \mathbf{P}(A).$$

For (2.5.3), let $B_1, B_2, \ldots \in \mathcal{J}$ with $\bigcup_n B_n \in \mathcal{J}$. Set $D_1 = B_1$, and $D_n = B_n \cap B_1^C \cap \ldots \cap B_{n-1}^C$ for $n \geq 2$. Then $\{D_n\}$ are disjoint, with $\bigcup_n D_n = \bigcup_n B_n$. Furthermore, since $\mathcal{J}$ is a semialgebra, each D_n can be written as a finite disjoint union of elements of $\mathcal{J}$, say $D_n = J_{n1} \dot{\cup} \ldots \dot{\cup} J_{nk_n}$. It then follows from (2.5.5) that

$$\mathbf{P}\left(\bigcup_n B_n\right) = \mathbf{P}\left(\bigcup_n D_n\right) = \mathbf{P}\left(\bigcup_n \bigcup_{i=1}^{k_n} J_{ni}\right) = \sum_n \sum_{i=1}^{k_n} \mathbf{P}(J_{ni}).$$

On the other hand,

$$B_n = \bigcup_{m \leq n} \bigcup_{i=1}^{k_m} (J_{mi} \cap B_n)$$

and the union is disjoint, with $J_{ni} \subseteq B_n$, so

$$\mathbf{P}(B_n) = \sum_{m \leq n} \sum_{i=1}^{k_m} \mathbf{P}(J_{mi} \cap B_n) \geq \sum_{i=1}^{k_n} \mathbf{P}(J_{ni} \cap B_n) = \sum_{i=1}^{k_n} \mathbf{P}(J_{ni}),$$

and hence

$$\sum_n \mathbf{P}(B_n) \geq \sum_n \sum_{i=1}^{k_n} \mathbf{P}(J_{ni}) = \mathbf{P}\left(\bigcup_n B_n\right). \qquad \blacksquare$$

Exercise 2.5.6. Suppose $\mathbf{P}$ satisfies (2.5.5) for *finite* collections $\{D_n\}$. Suppose further that, whenever $A_1, A_2, \ldots$ are finite unions of elements of $\mathcal{J}$ such that $A_{n+1} \subseteq A_n$ and $\bigcap_{n=1}^{\infty} A_n = \emptyset$, we have $\lim_{n \to \infty} \mathbf{P}(A_n) = 0$. Prove that $\mathbf{P}$ also satisfies (2.5.5) for countable collections $\{D_n\}$. [Hint: Set $A_n = \left(\bigcup_{j=1}^{\infty} D_j\right) \setminus \left(\bigcup_{i=1}^{n} D_i\right)$.]

The extension of Theorem 2.3.1 also has a *uniqueness* property:

Proposition 2.5.7. Let $\mathcal{J}$, $\mathbf{P}$, $\mathbf{P}^*$, and $\mathcal{M}$ be as in Theorem 2.3.1 (or as in Corollary 2.5.1 or 2.5.4). Let $\mathcal{F}$ be any σ-algebra with $\mathcal{J} \subseteq \mathcal{F} \subseteq \mathcal{M}$ (e.g.

$\mathcal{F} = \mathcal{M}$, or $\mathcal{F} = \sigma(\mathcal{J})$). Let $\mathbf{Q}$ be any probability measure on $\mathcal{F}$, such that $\mathbf{Q}(A) = \mathbf{P}(A)$ for all $A \in \mathcal{J}$. Then $\mathbf{Q}(A) = \mathbf{P}^*(A)$ for all $A \in \mathcal{F}$.

Proof.　For $A \in \mathcal{F}$, we compute

$$
\begin{aligned}
\mathbf{P}^*(A) &= \inf_{\substack{A_1, A_2, \dots \in \mathcal{J} \\ A \subseteq \bigcup_i A_i}} \sum_i \mathbf{P}(A_i) && \text{from (2.3.4)} \\
&= \inf_{\substack{A_1, A_2, \dots \in \mathcal{J} \\ A \subseteq \bigcup_i A_i}} \sum_i \mathbf{Q}(A_i) && \text{since } \mathbf{Q} = \mathbf{P} \text{ on } \mathcal{J} \\
&\geq \inf_{\substack{A_1, A_2, \dots \in \mathcal{J} \\ A \subseteq \bigcup_i A_i}} \mathbf{Q}(\textstyle\bigcup A_i) && \text{by countable subadditivity} \\
&\geq \inf_{\substack{A_1, A_2, \dots \in \mathcal{J} \\ A \subseteq \bigcup_i A_i}} \mathbf{Q}(A) && \text{by monotonicity} \\
&= \mathbf{Q}(A),
\end{aligned}
$$

i.e. $\mathbf{P}^*(A) \geq \mathbf{Q}(A)$. Similarly, $\mathbf{P}^*(A^C) \geq \mathbf{Q}(A^C)$, and then (2.1.1) implies $1 - \mathbf{P}^*(A) \geq 1 - \mathbf{Q}(A)$, i.e. $\mathbf{P}^*(A) \leq \mathbf{Q}(A)$. Hence, $\mathbf{P}^*(A) = \mathbf{Q}(A)$.　∎

Proposition 2.5.7 immediately implies the following:

Proposition 2.5.8.　Let $\mathcal{J}$ be a semialgebra of subsets of Ω, and let $\mathcal{F} = \sigma(\mathcal{J})$ be the generated σ-algebra. Let $\mathbf{P}$ and $\mathbf{Q}$ be two probability distributions defined on $\mathcal{F}$. Suppose that $\mathbf{P}(A) = \mathbf{Q}(A)$ for all $A \in \mathcal{J}$. Then $\mathbf{P} = \mathbf{Q}$, i.e. $\mathbf{P}(A) = \mathbf{Q}(A)$ for all $A \in \mathcal{F}$.

Proof.　Since $\mathbf{P}$ and $\mathbf{Q}$ are probability measures, they both satisfy (2.3.2) and (2.3.3). Hence, by Proposition 2.5.7, each of $\mathbf{P}$ and $\mathbf{Q}$ is equal to the $\mathbf{P}^*$ of Theorem 2.3.1.　∎

One useful special case of Proposition 2.5.8 is:

Corollary 2.5.9.　Let $\mathbf{P}$ and $\mathbf{Q}$ be two probability distributions defined on the collection $\mathcal{B}$ of Borel subsets of $\mathbf{R}$. Suppose $\mathbf{P}\big((-\infty, x]\big) = \mathbf{Q}\big((-\infty, x]\big)$ for all $x \in \mathbf{R}$. Then $\mathbf{P}(A) = \mathbf{Q}(A)$ for all $A \in \mathcal{B}$.

Proof.　Since $\mathbf{P}\big((y, \infty)\big) = 1 - \mathbf{P}\big((-\infty, y]\big)$, and $\mathbf{P}\big((x, y]\big) = 1 - \mathbf{P}\big((-\infty, x]\big) - \mathbf{P}\big((y, \infty)\big)$, and similarly for $\mathbf{Q}$, it follows that $\mathbf{P}$ and $\mathbf{Q}$ agree on

$$
\mathcal{J} = \big\{(-\infty, x] : x \in \mathbf{R}\big\} \cup \big\{(y, \infty) : y \in \mathbf{R}\big\} \cup \big\{(x, y] : x, y \in \mathbf{R}\big\} \cup \big\{\emptyset, \mathbf{R}\big\}. \tag{2.5.10}
$$

But $\mathcal{J}$ is a semialgebra (Exercise 2.7.10), and it follows from Exercise 2.4.5 that $\sigma(\mathcal{J}) = \mathcal{B}$. Hence, the result follows from Proposition 2.5.8.　∎

2.6. Coin tossing and other measures.

Now that we have Theorem 2.3.1 to help us, we can easily construct other probability triples as well.

For example, of frequent mention in probability theory is (independent, fair) coin tossing. To model the flipping of n coins, we can simply take $\Omega = \{(r_1, r_2, \ldots, r_n);\ r_i = 0 \text{ or } 1\}$ (where 0 stands for tails and 1 stands for heads), let $\mathcal{F} = 2^\Omega$ be the collection of all subsets of Ω, and define $\mathbf{P}$ by $\mathbf{P}(A) = |A|/2^n$ for $A \subseteq \mathcal{F}$. This is another example of a discrete probability space; and we know from Theorem 2.2.1 that these spaces present no difficulties.

But suppose now that we wish to model the flipping of a (countably) *infinite* number of coins. In this case we can let

$$\Omega = \{(r_1, r_2, r_3, \ldots);\ r_i = 0 \text{ or } 1\}$$

be the collection of all binary sequences. But what about $\mathcal{F}$ and $\mathbf{P}$?

Well, for each $n \in \mathbf{N}$ and each $a_1, a_2, \ldots, a_n \in \{0, 1\}$, let us define subsets $A_{a_1 a_2 \ldots a_n} \subseteq \Omega$ by

$$A_{a_1 a_2 \ldots a_n} = \{(r_1, r_2, \ldots) \in \Omega;\ r_i = a_i \text{ for } 1 \le i \le n\}.$$

(Thus, A_0 is the event that the first coin comes up tails; A_{11} is the event that the first two coins both come up heads; and A_{101} is the event that the first and third coins are heads while the second coin is tails.) Then we clearly want $\mathbf{P}(A_{a_1 a_2 \ldots a_n}) = 1/2^n$ for each set $A_{a_1 a_2 \ldots a_n}$. Hence, if we set

$$\mathcal{J} = \{A_{a_1 a_2 \ldots a_n};\ n \in \mathbf{N},\ a_1, a_2, \ldots, a_n \in \{0, 1\}\} \cup \{\emptyset, \Omega\},$$

then we already know how to define $\mathbf{P}(A)$ for each $A \in \mathcal{J}$. To apply the Extension Theorem (in this case, Corollary 2.5.4), we need to verify that certain conditions are satisfied.

Exercise 2.6.1. (a) Verify that the above $\mathcal{J}$ is a semialgebra.
(b) Verify that the above $\mathcal{J}$ and $\mathbf{P}$ satisfy (2.5.5) for *finite* collections $\{D_n\}$. [Hint: For a *finite* collection $\{D_n\} \subseteq \mathcal{J}$, there is $k \in \mathbf{N}$ such that the results of only coins 1 through k are specified by any D_n. Partition Ω into the corresponding 2^k subsets.]

Verifying (2.5.5) for countable collections unfortunately requires a bit of topology; the proof of this next lemma may be skipped.

Lemma 2.6.2. *The above $\mathcal{J}$ and $\mathbf{P}$ (for infinite coin tossing) satisfy (2.5.5).*

Proof (optional). In light of Exercises 2.6.1 and 2.5.6, it suffices to show that if $A_1, A_2, \ldots$ are finite unions of elements of $\mathcal{J}$ with $A_{n+1} \subseteq A_n$ and $\bigcap_{n=1}^{\infty} A_n = \emptyset$, then $\lim_{n \to \infty} \mathbf{P}(A_n) = 0$.

Give $\{0, 1\}$ the discrete topology, and give $\Omega = \{0, 1\} \times \{0, 1\} \times \cdots$ the corresponding product topology. Then Ω is a product of compact sets $\{0, 1\}$, and hence is itself compact by Tychonov's Theorem. Furthermore each element of $\mathcal{J}$ is a closed subset of Ω, since its complement is open in the product topology. Hence, each A_n is a closed subset of a compact space, and is therefore compact.

The *finite intersection property* of compact sets then implies that there is $N \in \mathbf{N}$ with $A_n = \emptyset$ for all $n > N$. In particular, $\mathbf{P}(A_n) \to 0$. ∎

Now that these conditions have been verified, it then follows from Corollary 2.5.4 that the probabilities for the special sets $A_{a_1 a_2 \ldots a_n} \in \mathcal{J}$ can automatically be extended to a σ-algebra $\mathcal{M}$ containing $\mathcal{J}$. (Once again, this σ-algebra will be quite complicated, and we will never understand it completely. But it is still essential mathematically that we know it exists.) This will be our probability triple for *infinite fair coin tossing*.

As a sample calculation, let $H_n = \{(r_1, r_2, \ldots) \in \Omega; r_n = 1\}$ be the event that the nth coin comes up heads. We certainly would hope that $H_n \in \mathcal{M}$, with $\mathbf{P}(H_n) = \frac{1}{2}$. Happily, this is indeed the case. We note that

$$H_n = \dot{\bigcup}_{r_1, r_2, \ldots, r_{n-1} \in \{0,1\}} A_{r_1 r_2 \ldots r_{n-1} 1},$$

the union being disjoint. Hence, since $\mathcal{M}$ is closed under countable (including finite) unions, we have $H_n \in \mathcal{M}$. Then, by countable additivity,

$$\mathbf{P}(H_n) = \sum_{r_1, r_2, \ldots, r_{n-1} \in \{0,1\}} \mathbf{P}(A_{r_1 r_2 \ldots r_{n-1} 1})$$

$$= \sum_{r_1, r_2, \ldots, r_{n-1} \in \{0,1\}} 1/2^n = 2^{n-1}/2^n = 1/2.$$

Remark. In fact, if we identify an element $x \in [0, 1]$ by its binary expansion $(r_1, r_2, \ldots)$, i.e. so that $x = \sum_{k=1}^{\infty} r_k/2^k$, then we see that in fact infinite fair coin tossing may be viewed as being "essentially" the same thing as Lebesgue measure on $[0, 1]$.

Next, given any two probability triples $(\Omega_1, \mathcal{F}_1, \mathbf{P}_1)$ and $(\Omega_2, \mathcal{F}_2, \mathbf{P}_2)$, we can define their *product measure* $\mathbf{P}$ on the Cartesian product set $\Omega_1 \times \Omega_2 = \{(\omega_1, \omega_2) : \omega_i \in \Omega_i \ (i = 1, 2)\}$. We set

$$\mathcal{J} = \{A \times B ; \ A \in \mathcal{F}_1, \ B \in \mathcal{F}_2\}, \qquad (2.6.3)$$

and define $\mathbf{P}(A \times B) = \mathbf{P}_1(A)\,\mathbf{P}_2(B)$ for $A \times B \in \mathcal{J}$. (The elements of $\mathcal{J}$ are called *measurable rectangles*.)

Exercise 2.6.4.   Verify that the above $\mathcal{J}$ is a semialgebra, and that $\emptyset, \Omega \in \mathcal{J}$ with $\mathbf{P}(\emptyset) = 0$ and $\mathbf{P}(\Omega) = 1$.

We will show later (Exercise 4.5.15) that these $\mathcal{J}$ and $\mathbf{P}$ satisfy (2.5.5). Hence, by Corollary 2.5.4, we can extend $\mathbf{P}$ to a σ-algebra containing $\mathcal{J}$. The resulting probability triple is called the *product measure* of $(\Omega_1, \mathcal{F}_1, P_1)$ and $(\Omega_2, \mathcal{F}_2, P_2)$.

An important special case of product measure is Lebesgue measure in higher dimensions. For example, in dimension two, we define 2-dimensional Lebesgue measure on $[0,1] \times [0,1]$ to be the product of Lebesgue measure on $[0,1]$ with itself. This is a probability measure on $\Omega = [0,1] \times [0,1]$ with the property that

$$\mathbf{P}\Big([a,b] \times [c,d]\Big) \;=\; (b-a)(d-c)\,, \qquad 0 \le a \le b \le 1, \quad 0 \le c \le d \le 1.$$

It is thus a measure of *area* in two dimensions.

More generally, we can inductively define d-dimensional Lebesgue measure on $[0,1]^d$ for any $d \in \mathbf{N}$, by taking the product of Lebesgue measure on $[0,1]$ with $(d-1)$-dimensional Lebesgue measure on $[0,1]^{d-1}$. When $d = 3$, Lebesgue measure on $[0,1] \times [0,1] \times [0,1]$ is a measure of *volume*.

2.7. Exercises.

Exercise 2.7.1.   Let $\Omega = \{1,2,3,4\}$. Determine whether or not each of the following is a σ-algebra.

(a) $\mathcal{F}_1 = \Big\{\emptyset,\ \{1,2\},\ \{3,4\},\ \{1,2,3,4\}\Big\}$.

(b) $\mathcal{F}_2 = \Big\{\emptyset,\ \{3\},\ \{4\},\ \{1,2\},\ \{3,4\},\ \{1,2,3\},\ \{1,2,4\},\ \{1,2,3,4\}\Big\}$.

(c) $\mathcal{F}_3 = \Big\{\emptyset,\ \{1,2\},\ \{1,3\},\ \{1,4\},\ \{2,3\},\ \{2,4\},\ \{3,4\},\ \{1,2,3,4\}\Big\}$.

Exercise 2.7.2.   Let $\Omega = \{1,2,3,4\}$, and let $\mathcal{J} = \{\{1\},\ \{2\}\}$. Describe explicitly the σ-algebra $\sigma(\mathcal{J})$ generated by $\mathcal{J}$.

Exercise 2.7.3.   Suppose $\mathcal{F}$ is a collection of subsets of Ω, such that $\Omega \in \mathcal{F}$.

(a) Suppose $\mathcal{F}$ is a algebra. Prove that $\mathcal{F}$ is a semialgebra.

(b) Suppose that whenever $A, B \in \mathcal{F}$, then also $A \setminus B \equiv A \cap B^C \in \mathcal{F}$. Prove that $\mathcal{F}$ is an algebra.

(c) Suppose that $\mathcal{F}$ is closed under complement, and also closed under finite *disjoint* unions (i.e. whenever $A, B \in \mathcal{F}$ are *disjoint*, then $A \cup B \in \mathcal{F}$). Give a counter-example to show that $\mathcal{F}$ might *not* be an algebra.

Exercise 2.7.4.   Let $\mathcal{F}_1, \mathcal{F}_2, \ldots$ be a sequence of collections of subsets of Ω, such that $\mathcal{F}_n \subseteq \mathcal{F}_{n+1}$ for each n.
(a) Suppose that each $\mathcal{F}_i$ is an algebra. Prove that $\bigcup_{i=1}^{\infty} \mathcal{F}_i$ is also an algebra.
(b) Suppose that each $\mathcal{F}_i$ is a σ-algebra. Show (by counter-example) that $\bigcup_{i=1}^{\infty} \mathcal{F}_i$ might *not* be a σ-algebra.

Exercise 2.7.5.   Suppose that $\Omega = \mathbf{N}$ is the set of positive integers, and $\mathcal{F}$ is the set of all subsets A such that either A or A^C is finite, and $\mathbf{P}$ is defined by $\mathbf{P}(A) = 0$ if A is finite, and $\mathbf{P}(A) = 1$ if A^C is finite.
(a) Is $\mathcal{F}$ an algebra?
(b) Is $\mathcal{F}$ a σ-algebra?
(c) Is $\mathbf{P}$ finitely additive?
(d) Is $\mathbf{P}$ countably additive on $\mathcal{F}$, meaning that if $A_1, A_2, \ldots \in \mathcal{F}$ are disjoint, and if it happens that $\bigcup_n A_n \in \mathcal{F}$, then $\mathbf{P}(\bigcup_n A_n) = \sum_n \mathbf{P}(A_n)$?

Exercise 2.7.6.   Suppose that $\Omega = [0, 1]$ is the unit interval, and $\mathcal{F}$ is the set of all subsets A such that either A or A^C is finite, and $\mathbf{P}$ is defined by $\mathbf{P}(A) = 0$ if A is finite, and $\mathbf{P}(A) = 1$ if A^C is finite.
(a) Is $\mathcal{F}$ an algebra?
(b) Is $\mathcal{F}$ a σ-algebra?
(c) Is $\mathbf{P}$ finitely additive?
(d) Is $\mathbf{P}$ countably additive on $\mathcal{F}$ (as in the previous exercise)?

Exercise 2.7.7.   Suppose that $\Omega = [0, 1]$ is the unit interval, and $\mathcal{F}$ is the set of all subsets A such that either A or A^C is countable (i.e., finite or countably infinite), and $\mathbf{P}$ is defined by $\mathbf{P}(A) = 0$ if A is countable, and $\mathbf{P}(A) = 1$ if A^C is countable.
(a) Is $\mathcal{F}$ an algebra?
(b) Is $\mathcal{F}$ a σ-algebra?
(c) Is $\mathbf{P}$ finitely additive?
(d) Is $\mathbf{P}$ countably additive on $\mathcal{F}$?

Exercise 2.7.8.   For the example of Exercise 2.7.7, is $\mathbf{P}$ *uncountably* additive (cf. page 2)?

Exercise 2.7.9.   Let $\mathcal{F}$ be a σ-algebra, and write $|\mathcal{F}|$ for the total number of subsets in $\mathcal{F}$. Prove that if $|\mathcal{F}| < \infty$ (i.e., if $\mathcal{F}$ consists of just a *finite* number of subsets), then $|\mathcal{F}| = 2^m$ for some $m \in \mathbf{N}$. [Hint: Consider those non-empty subsets in $\mathcal{F}$ which do not contain any other non-empty setset in $\mathcal{F}$. How can all subsets in $\mathcal{F}$ be "built up" from these particular subsets?]

Exercise 2.7.10. Prove that the collection $\mathcal{J}$ of (2.5.10) is a semialgebra.

Exercise 2.7.11. Let $\Omega = [0, 1]$. Let $\mathcal{J}'$ be the set of all half-open intervals of the form $(a, b]$, for $0 \le a < b \le 1$, together with the sets $\emptyset$, Ω, and $\{0\}$.
(a) Prove that $\mathcal{J}'$ is a semialgebra.
(b) Prove that $\sigma(\mathcal{J}') = \mathcal{B}$, i.e. that the σ-algebra generated by this $\mathcal{J}'$ is equal to the σ-algebra generated by the $\mathcal{J}$ of (2.4.1).
(c) Let $\mathcal{B}_0'$ be the collection of all finite disjoint unions of elements of $\mathcal{J}'$. Prove that $\mathcal{B}_0'$ is an algebra. Is $\mathcal{B}_0'$ the same as the algebra $\mathcal{B}_0$ defined in (2.2.4)?
[Remark: Some treatments of Lebesgue measure use $\mathcal{J}'$ instead of $\mathcal{J}$.]

Exercise 2.7.12. Let K be the Cantor set as defined in Subsection 2.4. Let $D_n = K \oplus \frac{1}{n}$ where $K \oplus \frac{1}{n}$ is defined as in (1.2.4). Let $B = \bigcup_{n=1}^{\infty} D_n$.
(a) Draw a rough sketch of D_3.
(b) What is $\lambda(D_3)$?
(c) Draw a rough sketch of B.
(d) What is $\lambda(B)$?

Exercise 2.7.13. Give an example of a sample space Ω, a semialgebra $\mathcal{J}$, and a non-negative function $\mathbf{P} : \mathcal{J} \to \mathbf{R}$ with $\mathbf{P}(\emptyset) = 0$ and $\mathbf{P}(\Omega) = 1$, such that (2.5.5) is *not* satisfied.

Exercise 2.7.14. Let $\Omega = \{1, 2, 3, 4\}$, with $\mathcal{F}$ the collection of all subsets of Ω. Let $\mathbf{P}$ and $\mathbf{Q}$ be two probability measures on $\mathcal{F}$, such that $\mathbf{P}\{1\} = \mathbf{P}\{2\} = \mathbf{P}\{3\} = \mathbf{P}\{4\} = 1/4$, and $\mathbf{Q}\{2\} = \mathbf{Q}\{4\} = 1/2$, extended to $\mathcal{F}$ by linearity. Finally, let $\mathcal{J} = \{\emptyset, \Omega, \{1, 2\}, \{2, 3\}, \{3, 4\}, \{1, 4\}\}$.
(a) Prove that $\mathbf{P}(A) = \mathbf{Q}(A)$ for all $A \in \mathcal{J}$.
(b) Prove that there is $A \in \sigma(\mathcal{J})$ with $\mathbf{P}(A) \ne \mathbf{Q}(A)$.
(c) Why does this not contradict Proposition 2.5.8?

Exercise 2.7.15. Let $(\Omega, \mathcal{M}, \lambda)$ be Lebesgue measure on the interval $[0, 1]$. Let
$$\Omega' = \{(x, y) \in \mathbf{R}^2 ; \ 0 < x \le 1, \ 0 < y \le 1\} \, .$$
Let $\mathcal{F}$ be the collection of all subsets of Ω' of the form
$$\{(x, y) \in \mathbf{R}^2 ; \ x \in A, \ 0 < y \le 1\}$$
for some $A \in \mathcal{M}$. Finally, define a probability $\mathbf{P}$ on $\mathcal{F}$ by
$$\mathbf{P}\left(\{(x, y) \in \mathbf{R}^2 ; \ x \in A, \ 0 < y \le 1\}\right) = \lambda(A) \, .$$

(a) Prove that $(\Omega', \mathcal{F}, \mathbf{P})$ is a probability triple.
(b) Let $\mathbf{P}^*$ be the outer measure corresponding to $\mathbf{P}$ and $\mathcal{F}$. Define the subset $S \subseteq \Omega'$ by

$$S = \left\{ (x,y) \in \mathbf{R}^2 \,;\, 0 < x \leq 1, \ y = 1/2 \right\}.$$

(Note that $S \notin \mathcal{F}$.) Prove that $\mathbf{P}^*(S) = 1$ and $\mathbf{P}^*(S^C) = 1$.

Exercise 2.7.16.   (a) Where in the proof of Theorem 2.3.1 was assumption (2.3.3) used?
(b) How would the conclusion of Theorem 2.3.1 by modified if assumption (2.3.3) were dropped (but all other assumptions remained the same)?

Exercise 2.7.17.   Let $\Omega = \{1,2\}$, and let $\mathcal{J}$ be the collection of all subsets of Ω, with $P(\emptyset) = 0$, $P(\Omega) = 1$, and $\mathbf{P}\{1\} = \mathbf{P}\{2\} = 1/3$.
(a) Verify that all assumptions of Theorem 2.3.1 other than (2.3.3) are satisfied.
(b) Verify that assumption (2.3.3) is not satisfied.
(c) Describe precisely the $\mathcal{M}$ and $\mathbf{P}^*$ that would result in this example from the modified version of Theorem 2.3.1 in Exercise 2.7.16(b).

Exercise 2.7.18.   Let $\Omega = \{1,2\}$, $\mathcal{J} = \{\emptyset, \Omega, \{1\}\}$, $P(\emptyset) = 0$, $P(\Omega) = 1$, and $P(\{1\}) = 1/3$.
(a) Can Theorem 2.3.1, Corollary 2.5.1, or Corollary 2.5.4 be applied in this case? Why or why not?
(b) Can this $\mathbf{P}$ be extended to a valid probability measure? Explain.

Exercise 2.7.19.   Let Ω be a finite non-empty set, and let $\mathcal{J}$ consist of all singletons in Ω, together with $\emptyset$ and Ω. Let $p : \Omega \to [0,1]$ with $\sum_{\omega \in \Omega} p(\omega) = 1$, and define $\mathbf{P}(\emptyset) = 0$, $\mathbf{P}(\Omega) = 1$, and $\mathbf{P}\{\omega\} = p(\omega)$ for all $\omega \in \Omega$.
(a) Prove that $\mathcal{J}$ is a semialgebra.
(b) Prove that (2.3.2) and (2.3.3) are satisfied.
(c) Describe precisely the $\mathcal{M}$ and $\mathbf{P}^*$ that result from applying Theorem 2.3.1.
(d) Are these $\mathcal{M}$ and $\mathbf{P}^*$ the same as those described in Theorem 2.2.1?

Exercise 2.7.20.   Let $\mathbf{P}$ and $\mathbf{Q}$ be two probability measures defined on the same sample space Ω and σ-algebra $\mathcal{F}$.
(a) Suppose that $\mathbf{P}(A) = \mathbf{Q}(A)$ for all $A \in \mathcal{F}$ with $\mathbf{P}(A) \leq \frac{1}{2}$. Prove that $\mathbf{P} = \mathbf{Q}$, i.e. that $\mathbf{P}(A) = \mathbf{Q}(A)$ for all $A \in \mathcal{F}$.

(b) Give an example where $\mathbf{P}(A) = \mathbf{Q}(A)$ for all $A \in \mathcal{F}$ with $\mathbf{P}(A) < \frac{1}{2}$, but such that $\mathbf{P} \neq \mathbf{Q}$, i.e. that $\mathbf{P}(A) \neq \mathbf{Q}(A)$ for some $A \in \mathcal{F}$.

Exercise 2.7.21. Let λ be Lebesgue measure in dimension two, i.e. Lebesgue measure on $[0,1] \times [0,1]$. Let A be the triangle $\{(x,y) \in [0,1] \times [0,1] \,; \, y < x\}$. Prove that A is measurable with respect to λ, and compute $\lambda(A)$.

Exercise 2.7.22. Let $(\Omega_1, \mathcal{F}_1, \mathbf{P}_1)$ be Lebesgue measure on $[0,1]$. Consider a second probability triple, $(\Omega_2, \mathcal{F}_2, \mathbf{P}_2)$, defined as follows: $\Omega_2 = \{1,2\}$, $\mathcal{F}_2$ consists of all subsets of Ω_2, and $\mathbf{P}_2$ is defined by $\mathbf{P}_2\{1\} = \frac{1}{3}$, $\mathbf{P}_2\{2\} = \frac{2}{3}$, and additivity. Let $(\Omega, \mathcal{F}, \mathbf{P})$ be the product measure of $(\Omega_1, \mathcal{F}_1, \mathbf{P}_1)$ and $(\Omega_2, \mathcal{F}_2, \mathbf{P}_2)$.
(a) Express each of Ω, $\mathcal{F}$, and $\mathbf{P}$ as explicitly as possible.
(b) Find a set $A \in \mathcal{F}$ such that $\mathbf{P}(A) = \frac{3}{4}$.

2.8. Section summary.

The section gave a formal definition of a probability triple $(\Omega, \mathcal{F}, \mathbf{P})$, consisting of a sample space Ω, a σ-algebra $\mathcal{F}$, and a probability measure $\mathbf{P}$, and derived certain basic properties of them. It then considered the question of how to *construct* such probability triples. Discrete spaces (with countable Ω) were straightforward, but other spaces were more challenging. The key tool was the Extension Theorem, which said that once a probability measure has been constructed on a semialgebra, it can then automatically be extended to a σ-algebra.

The Extension Theorem allowed us to construct Lebesgue measure on $[0,1]$, and to consider some of its basic properties. It also allowed us to construct other probability triples such as infinite coin tossing, product measures, and multi-dimensional Lebesgue measure.

3. Further probabilistic foundations.

Now that we understand probability triples well, we discuss some additional essential ingredients of probability theory. Throughout Section 3 (and, indeed, throughout most of this text and most of probability theory in general), we shall assume that there is an underlying probability triple $(\Omega, \mathcal{F}, \mathbf{P})$ with respect to which all further probability objects are defined. This assumption shall be so universal that we will often not even mention it.

3.1. Random variables.

If we think of a sample space Ω as the set of all possible random outcomes of some experiment, then a *random variable* assigns a numerical value to each of these outcomes. More formally, we have

Definition 3.1.1. Given a probability triple $(\Omega, \mathcal{F}, \mathbf{P})$, a *random variable* is a function X from Ω to the real numbers $\mathbf{R}$, such that

$$\{\omega \in \Omega \,;\, X(w) \le x\} \in \mathcal{F}, \qquad x \in \mathbf{R}. \qquad (3.1.2)$$

Equation (3.1.2) is a technical requirement, and states that the function X must be *measurable*. It can also be written as $\{X \le x\} \in \mathcal{F}$, or $X^{-1}((-\infty, x]) \in \mathcal{F}$, for all $x \in \mathbf{R}$. Since complements and unions and intersections are preserved under inverse images (see Subsection A.1), it follows from Exercise 2.4.5 that equation (3.1.2) is equivalent to saying that $X^{-1}(B) \in \mathcal{F}$ for every Borel set B. That is, the set $X^{-1}(B)$, also written $\{X \in B\}$, is indeed an event. So, for any Borel set B, it makes sense to talk about $\mathbf{P}(X \in B)$, the probability that X lies in B.

Example 3.1.3. Suppose that $(\Omega, \mathcal{F}, \mathbf{P})$ is Lebesgue measure on $[0, 1]$, then we might define some random variables X, Y, and Z by $X(\omega) = \omega$, $Y(\omega) = 2\omega$, and $Z(\omega) = 3\omega + 4$. We then have, for example, that $Y = 2X$, and $Z = 3X + 4 = \frac{3}{2}Y + 4$. Also, $\mathbf{P}(Y \le 1/3) = \mathbf{P}\{\omega \,;\, Y(\omega) < 1/3\} = \mathbf{P}\{\omega \,;\, 2\omega < 1/3\} = \mathbf{P}([0, 1/6]) = 1/6$.

Exercise 3.1.4. For Example 3.1.3, compute $\mathbf{P}(Z > a)$ and $\mathbf{P}(X < a \text{ \underline{and} } Y < b)$ as functions of $a, b \in \mathbf{R}$.

Now, not all functions from Ω to $\mathbf{R}$ are random variables. For example, let $(\Omega, \mathcal{F}, \mathbf{P})$ be Lebesgue measure on $[0, 1]$, and let $H \subset \Omega$ be the non-measurable set of Proposition 2.4.8. Define $X : \Omega \to \mathbf{R}$ by $X = 1_{H^c}$, so

$X(\omega) = 0$ for $\omega \in H$, and $X(\omega) = 1$ for $\omega \notin H$. Then $\{\omega \in \Omega : X(\omega) \leq 1/2\} = H \notin \mathcal{F}$, so X is not a random variable.

On the other hand, the following proposition shows that condition (3.1.2) is preserved under usual arithmetic and limits. In practice, this means that if functions from Ω to $\mathbf{R}$ are constructed in "usual" ways, then (3.1.2) will be satisfied, so the functions will indeed be random variables.

Proposition 3.1.5.   (i) If $X = 1_A$ is the indicator of some event $A \in \mathcal{F}$, then X is a random variable.
(ii) If X and Y are random variables and $c \in \mathbf{R}$, then $X+c$, cX, X^2, $X+Y$, and XY are all random variables.
(iii) If $Z_1, Z_2, \ldots$ are random variables such that $\lim_{n \to \infty} Z_n(\omega)$ exists for each $\omega \in \Omega$, and $Z(\omega) = \lim_{n \to \infty} Z_n(\omega)$, then Z is also a random variable.

Proof.   (i) If $X = 1_A$ for $A \in \mathcal{F}$, then $X^{-1}(B)$ must be one of A, A^C, $\emptyset$, or Ω, so $X^{-1}(B) \in \mathcal{F}$.
(ii) The first two of these assertions are immediate. The third follows since for $y \geq 0$, $\{X^2 \leq y\} = \{X \in [-\sqrt{y}, \sqrt{y}]\} \in \mathcal{F}$. For the fourth, note (by finding a rational number $r \in (X, x - Y)$) that

$$\{X + Y < x\} = \bigcup_{r \text{ rational}} (\{X < r\} \cap \{Y < x - r\}) \in \mathcal{F}.$$

The fifth assertion then follows since $XY = \frac{1}{2}\left[(X + Y)^2 - X^2 - Y^2\right]$.
(iii) For $x \in \mathbf{R}$,

$$\{Z \leq x\} = \bigcap_{m=1}^{\infty} \bigcup_{n=1}^{\infty} \bigcap_{k=n}^{\infty} \left\{Z_k \leq x + \frac{1}{m}\right\}. \tag{3.1.6}$$

But Z_k is a random variable, so $\{Z_k \leq x + \frac{1}{m}\} \in \mathcal{F}$. Then, since $\mathcal{F}$ is a σ-algebra, we must have $\{Z \leq x\} \in \mathcal{F}$. ∎

Exercise 3.1.7.   Prove (3.1.6). [Hint: remember the definition of $X(\omega) = \lim_{n \to \infty} X_n(\omega)$, cf. Subsection A.3.]

Suppose now that X is a random variable, and $f : \mathbf{R} \to \mathbf{R}$ is a function from $\mathbf{R}$ to $\mathbf{R}$ which is *Borel-measurable*, meaning that $f^{-1}(A) \in \mathcal{B}$ for any $A \in \mathcal{B}$ (where $\mathcal{B}$ is the collection of Borel sets of $\mathbf{R}$). (Equivalently, f is a random variable corresponding to $\Omega = \mathbf{R}$ and $\mathcal{F} = \mathcal{B}$.) We can define a new random variable $f(X)$, the *composition* of X with f, by $f(X)(\omega) = f(X(\omega))$ for each $\omega \in \Omega$. Then (3.1.2) is satisfied since for $B \in \mathcal{B}$, $\{f(X) \in B\} = \{X \in f^{-1}(B)\} \in \mathcal{F}$.

Proposition 3.1.8. *If f is a continuous function, or a piecewise-continuous function, then f is Borel-measurable.*

Proof. A basic result of point-set topology says that if f is continuous, then $f^{-1}(O)$ is an open subset of $\mathbf{R}$ whenever O is. In particular, $f^{-1}((x,\infty))$ is open, so $f^{-1}((x,\infty)) \in \mathcal{B}$, so $f^{-1}((-\infty,x]) \in \mathcal{B}$.

If f is piecewise-continuous, then we can write $f = f_1 \mathbf{1}_{I_1} + f_2 \mathbf{1}_{I_2} + \ldots + f_n \mathbf{1}_{I_n}$ where the f_j are continuous and the $\{I_j\}$ are disjoint intervals. It follows from the above and Proposition 3.1.5 that f is Borel-measurable. ∎

For example, if $f(x) = x^k$ for $k \in \mathbf{N}$, then f is Borel-measurable. Hence, if X is a random variable, then so is X^k for all $k \in \mathbf{N}$.

Remark 3.1.9. In probability theory, the underlying probability triple $(\Omega, \mathcal{F}, \mathbf{P})$ is usually *complete* (cf. Exercise 2.3.16; for example this is always true for discrete probability spaces, or for those such as Lebesgue measure constructed using the Extension Theorem). In that case, if X is a random variable, and $Y : \Omega \to \mathbf{R}$ such that $\mathbf{P}(X = Y) = 1$, then Y must also be a random variable.

Remark 3.1.10. In Definition 3.1.1, we assume that X is a *real-valued* random variable, i.e. that it maps Ω into the set of real numbers equipped with the Borel σ-algebra. More generally, one could consider a random variable which mapped Ω to an arbitrary second *measurable space*, i.e. to some second non-empty set Ω' with its own collection $\mathcal{F}'$ of measurable subsets. We would then have $X : \Omega \to \Omega'$, with condition (3.1.2) replaced by the condition that $X^{-1}(A') \in \mathcal{F}$ whenever $A' \in \mathcal{F}'$.

3.2. Independence.

Informally, events or random variables are *independent* if they do not affect each other's probabilities. Thus, two events A and B are independent if $\mathbf{P}(A \cap B) = \mathbf{P}(A)\mathbf{P}(B)$. Intuitively, the probabilistic proportion of the event B which also includes A (i.e., $\mathbf{P}(A \cap B)/\mathbf{P}(B)$) is equal to the overall probability of A (i.e., $\mathbf{P}(A)$) – the definition uses products to avoid division by zero.

Three events A, B, and C are said to be independent if *all* of the following equations are satisfied: $\mathbf{P}(A \cap B) = \mathbf{P}(A)\mathbf{P}(B)$; $\mathbf{P}(A \cap C) = \mathbf{P}(A)\mathbf{P}(C)$; $\mathbf{P}(B \cap C) = \mathbf{P}(B)\mathbf{P}(C)$; and $\mathbf{P}(A \cap B \cap C) = \mathbf{P}(A)\mathbf{P}(B)\mathbf{P}(C)$. It is *not* sufficient (see Exercise 3.6.3) to check just the final – or just the first three – of these equations. More generally, a possibly-infinite collection $\{A_\alpha\}_{\alpha \in I}$

of events is said to be independent if for each $j \in \mathbf{N}$ and each distinct *finite* choice $\alpha_1, \alpha_2, \ldots, \alpha_j \in I$, we have

$$\mathbf{P}(A_{\alpha_1} \cap A_{\alpha_2} \cap \ldots \cap A_{\alpha_j}) = \mathbf{P}(A_{\alpha_1})\mathbf{P}(A_{\alpha_2})\ldots\mathbf{P}(A_{\alpha_j}). \qquad (3.2.1)$$

Exercise 3.2.2.   Suppose (3.2.1) is satisfied.
(a) Show that (3.2.1) is still satisfied if A_{α_1} is replaced by $A_{\alpha_1}^C$.
(b) Show that (3.2.1) is still satisfied if each A_{α_i} is replaced by the corresponding $A_{\alpha_i}^C$.
(c) Prove that if $\{A_\alpha\}_{\alpha \in I}$ is independent, then so is $\{A_\alpha^C\}_{\alpha \in I}$.

We shall on occasion also talk about independence of *collections* of events. Collections of events $\{\mathcal{A}_\alpha; \alpha \in I\}$ are independent if for all $j \in \mathbf{N}$, for all distinct $\alpha_1, \ldots, \alpha_j \in I$, and for all $A_1 \in \mathcal{A}_{\alpha_1}, \ldots, A_j \in \mathcal{A}_{\alpha_j}$, equation (3.2.1) holds.

We shall also talk about independence of random variables. Random variables X and Y are independent if for all Borel sets S_1 and S_2, the events $X^{-1}(S_1)$ and $Y^{-1}(S_2)$ are independent, i.e. $P(X \in S_1, Y \in S_2) = \mathbf{P}(X \in S_1)\,\mathbf{P}(Y \in S_2)$. More generally, a collection $\{X_\alpha; \alpha \in I\}$ of random variables are independent if for all $j \in \mathbf{N}$, for all distinct $\alpha_1, \ldots, \alpha_j \in I$, and for all Borel sets $S_1, \ldots, S_j$, we have

$$\mathbf{P}\left(X_{\alpha_1} \in S_1,\ X_{\alpha_2} \in S_2,\ \ldots, X_{\alpha_j} \in S_j\right)$$

$$= \mathbf{P}(X_{\alpha_1} \in S_1)\,\mathbf{P}(X_{\alpha_2} \in S_2)\ \ldots\ \mathbf{P}(X_{\alpha_j} \in S_j).$$

Independence is preserved under deterministic transformations:

Proposition 3.2.3.   *Let X and Y be independent random variables. Let $f, g : \mathbf{R} \to \mathbf{R}$ be Borel-measurable functions. Then the random variables $f(X)$ and $g(Y)$ are independent.*

Proof.   For Borel $S_1, S_2 \subseteq \mathbf{R}$, we compute that

$$\mathbf{P}\left(f(X) \in S_1,\ g(Y) \in S_2\right) = \mathbf{P}\left(X \in f^{-1}(S_1),\ Y \in g^{-1}(S_2)\right)$$

$$= \mathbf{P}\left(X \in f^{-1}(S_1)\right)\,\mathbf{P}\left(Y \in g^{-1}(S_2)\right)$$

$$= \mathbf{P}\left(f(X) \in S_1\right)\,\mathbf{P}\left(g(Y) \in S_2\right). \qquad \blacksquare$$

We also have the following.

Proposition 3.2.4.   *Let X and Y be two random variables, defined jointly on some probability triple $(\Omega, \mathcal{F}, \mathbf{P})$. Then X and Y are independent if and only if $\mathbf{P}(X \leq x,\ Y \leq y) = \mathbf{P}(X \leq x)\,\mathbf{P}(Y \leq y)$ for all $x, y \in \mathbf{R}$.*

Proof. The "only if" part is immediate from the definition.

For the "if" part, fix $x \in \mathbf{R}$ with $\mathbf{P}(X \leq x) > 0$, and define the measure $\mathbf{Q}$ on the Borel subsets of $\mathbf{R}$ by $\mathbf{Q}(S) = \mathbf{P}(X \leq x, \, Y \in S) / \mathbf{P}(X \leq x)$. Then by assumption, $\mathbf{Q}((-\infty, y]) = \mathbf{P}(Y \leq y)$ for all $y \in \mathbf{R}$. It follows from Corollary 2.5.9 that $\mathbf{Q}(S) = \mathbf{P}(Y \in S)$ for all Borel $S \subseteq \mathbf{R}$, i.e. that $\mathbf{P}(X \leq x, \, Y \in S) = \mathbf{P}(X \leq x) \, \mathbf{P}(Y \in S)$.

Then, for fixed Borel $S \subseteq \mathbf{R}$, let $\mathbf{R}(T) = \mathbf{P}(X \in T, \, Y \in S) / \mathbf{P}(Y \in S)$. By the above, it follows that $\mathbf{R}((-\infty, x]) = \mathbf{P}(X \leq x)$ for each $x \in \mathbf{R}$. It then follows from Corollary 2.5.9 that $\mathbf{R}(T) = \mathbf{P}(X \in T)$ for all Borel $T \subseteq \mathbf{R}$, i.e. that $\mathbf{P}(X \in T, \, Y \in S) = \mathbf{P}(X \in T) \, \mathbf{P}(Y \in S)$ for all Borel $S, T \subseteq \mathbf{R}$. Hence, X and Y are independent. ∎

Independence will come up often in this text, and its significance will become more clear as we proceed.

3.3. Continuity of probabilities.

Given a probability triple $(\Omega, \mathcal{F}, \mathbf{P})$, and events $A, A_1, A_2, \ldots \in \mathcal{F}$, we write $\{A_n\} \nearrow A$ to mean that $A_1 \subseteq A_2 \subseteq A_3 \subseteq \ldots$, and $\bigcup_n A_n = A$. In words, the events A_n *increase* to A. Similarly, we write $\{A_n\} \searrow A$ to mean that $\{A_n^C\} \nearrow A^C$, or equivalently that $A_1 \supseteq A_2 \supseteq A_3 \supseteq \ldots$, and $\bigcap_n A_n = A$. In words, the events A_n *decrease* to A. We then have

Proposition 3.3.1. *(Continuity of probabilities.)* If $\{A_n\} \nearrow A$ or $\{A_n\} \searrow A$, then $\lim_{n \to \infty} \mathbf{P}(A_n) = \mathbf{P}(A)$.

Proof. Suppose $\{A_n\} \nearrow A$. Let $B_n = A_n \cap A_{n-1}^C$. Then the $\{B_n\}$ are disjoint, with $\bigcup B_n = \bigcup A_n = A$. Hence,

$$\mathbf{P}(A) = \mathbf{P}\left(\overset{\bullet}{\bigcup_m} B_m\right) = \sum_{m=1}^{\infty} \mathbf{P}(B_m)$$

$$= \lim_{n \to \infty} \sum_{m=1}^{n} \mathbf{P}(B_m) = \lim_{n \to \infty} \mathbf{P}\left(\overset{\bullet}{\underset{m \leq n}{\bigcup}} B_m\right) = \lim_{n \to \infty} \mathbf{P}(A_n)$$

(where the last equality is the only time we use that the $\{A_m\}$ are a nested sequence).

If instead $\{A_n\} \searrow A$, then $\{A_n^C\} \nearrow A^C$, so

$$\mathbf{P}(A) = 1 - \mathbf{P}(A^C) = 1 - \lim_n \mathbf{P}(A_n^C)$$

$$= \lim_n \left(1 - \mathbf{P}(A_n^C)\right) = \lim_n \mathbf{P}(A_n). \qquad \blacksquare$$

If the $\{A_n\}$ are *not* nested, then we may not have $\lim_n \mathbf{P}(A_n) = \mathbf{P}(A)$. For example, suppose that $A_n = \Omega$ for n odd, but $A_n = \emptyset$ for n even. Then $\mathbf{P}(A_n)$ alternates between 0 and 1, so that $\lim_n \mathbf{P}(A_n)$ does not exist.

3.4. Limit events.

Given events $A_1, A_2, \ldots \in \mathcal{F}$, we define

$$\limsup_n A_n = \{A_n \ i.o.\} = \bigcap_{n=1}^{\infty} \bigcup_{k=n}^{\infty} A_k$$

and

$$\liminf_n A_n = \{A_n \ a.a.\} = \bigcup_{n=1}^{\infty} \bigcap_{k=n}^{\infty} A_k \, .$$

The event $\limsup_n A_n$ is referred to as "A_n infinitely often"; it stands for those $\omega \in \Omega$ which are in infinitely many of the A_n. Intuitively, it is the event that infinitely many of the events A_n occur. Similarly, the event $\liminf_n A_n$ is referred to as "A_n almost always"; intuitively, it is the event that all but a finite number of the events A_n occur.

Since $\mathcal{F}$ is a σ-algebra, we see that $\limsup_n A_n \in \mathcal{F}$ and $\liminf_n A_n \in \mathcal{F}$. Also, by de Morgan's laws, $(\limsup_n A_n)^C = \liminf_n (A_n^C)$, so $\mathbf{P}(A_n \ i.o.) = 1 - \mathbf{P}(A_n^C \ a.a.)$.

For example, suppose $(\Omega, \mathcal{F}, \mathbf{P})$ is infinite fair coin tossing, and H_n is the event that the nth coin is heads. Then $\limsup_n H_n$ is the event that there are infinitely many heads. Also, $\liminf_n H_n$ is the event that all but a finite number of the coins were heads, i.e. that there were only finitely many tails.

Proposition 3.4.1.   *We always have*

$$\mathbf{P}\Big(\liminf_n A_n\Big) \leq \liminf_n \mathbf{P}(A_n) \leq \limsup_n \mathbf{P}(A_n) \leq \mathbf{P}\Big(\limsup_n A_n\Big).$$

Proof.   The middle inequality holds by definition, and the last inequality follows similarly to the first, so we prove only the first inequality. We note that as $n \to \infty$, the events $\{ \bigcap_{k=n}^{\infty} A_k \}$ are *increasing* (cf. page 33) to $\liminf_n A_n$. Hence, by continuity of probabilities,

$$\mathbf{P}\Big(\liminf_n A_n\Big) \equiv \mathbf{P}\Big(\bigcup_n \bigcap_{k=n}^{\infty} A_k\Big) = \lim_{n \to \infty} \mathbf{P}\Big(\bigcap_{k=n}^{\infty} A_k\Big)$$

$$= \liminf_{n \to \infty} \mathbf{P} \left(\bigcap_{k=n}^{\infty} A_k \right) \leq \liminf_{n \to \infty} \mathbf{P}(A_n),$$

where the final equality follows by definition (if a limit exists, then it is equal to the liminf), and the final inequality follows from monotonicity (2.1.2). ∎

For example, again considering infinite fair coin tossing, with H_n the event that the nth coin is heads. Proposition 3.4.1 says that $\mathbf{P}(H_n \; i.o.) \geq \frac{1}{2}$, which is interesting but vague. To improve this result, we require a more powerful theorem.

Theorem 3.4.2. *(The Borel-Cantelli Lemma.) Let $A_1, A_2, \ldots \in \mathcal{F}$.*
(i) If $\sum_n \mathbf{P}(A_n) < \infty$, then $\mathbf{P}(\limsup_n A_n) = 0$.
(ii) If $\sum_n \mathbf{P}(A_n) = \infty$, and $\{A_n\}$ are independent, then $\mathbf{P}(\limsup_n A_n) = 1$.

Proof. For (i), we note that for any $m \in \mathbf{N}$, we have by countable subadditivity that

$$\mathbf{P} \left(\limsup_n A_n \right) \leq \mathbf{P} \left(\bigcup_{k=m}^{\infty} A_k \right) \leq \sum_{k=m}^{\infty} \mathbf{P}(A_k),$$

which goes to 0 as $m \to \infty$ if the sum is convergent.

For (ii), since $(\limsup_n A_n)^C = \bigcup_{n=1}^{\infty} \bigcap_{k=n}^{\infty} A_k^C$, it suffices (by countable subadditivity) to show that $\mathbf{P} \left(\bigcap_{k=n}^{\infty} A_k^C \right) = 0$ for each $n \in \mathbf{N}$. Well, for $n, m \in \mathbf{N}$, we have by independence and Exercise 3.2.2 (and since $1 - x \leq e^{-x}$ for any real number x) that

$$
\begin{aligned}
\mathbf{P}(\bigcap_{k=n}^{\infty} A_k^C) &\leq \mathbf{P}(\bigcap_{k=n}^{n+m} A_k^C) \\
&= \prod_{k=n}^{n+m} \left(1 - \mathbf{P}(A_k) \right) \\
&\leq \prod_{k=n}^{n+m} e^{-\mathbf{P}(A_k)} \\
&= e^{-\sum_{k=n}^{n+m} \mathbf{P}(A_k)},
\end{aligned}
$$

which goes to 0 as $m \to \infty$ if the sum is divergent. ∎

This theorem is striking since it asserts that if $\{A_n\}$ are independent, then $\mathbf{P}(\limsup_n A_n)$ is always either 0 or 1 – it is never $\frac{1}{2}$ or $\frac{3}{4}$ or any other value. In the next section we shall see that this statement is true even more generally.

We note that the independence assumption for part (ii) of Theorem 3.4.2 cannot simply be omitted. For example, consider infinite fair coin tossing, and let $A_1 = A_2 = A_3 = \ldots = \{r_1 = 1\}$, i.e. let *all* the events be the event that the *first* coin comes up heads. Then the $\{A_n\}$ are clearly *not* independent. And, we clearly have $\mathbf{P}(\limsup_n A_n) = \mathbf{P}(r_1 = 1) = \frac{1}{2}$.

Theorem 3.4.2 provides very precise information about $\mathbf{P}(\limsup_n A_n)$ in many cases. Consider again infinite fair coin tossing, with H_n the event that the nth coin is heads. This theorem shows that $\mathbf{P}(H_n \ i.o.) = 1$, i.e. there is probability 1 that an infinite sequence of coins will contain infinitely many heads. Furthermore, $\mathbf{P}(H_n \ a.a.) = 1 - \mathbf{P}(H_n^C \ i.o.) = 1 - 1 = 0$, so the infinite sequence will never contain all but finitely many heads.

Similarly, we have that

$$\mathbf{P}\left\{H_{2^n+1} \cap H_{2^n+2} \cap \ldots \cap H_{2^n+[\log_2 n]} \ i.o.\right\} = 1\,,$$

since the events $\{H_{2^n+1} \cap H_{2^n+2} \cap \ldots \cap H_{2^n+[\log_2 n]}$ are seen to be independent for different values of n, and since their probabilities are approximately $1/n$ which sums to infinity. On the other hand,

$$\mathbf{P}\left\{H_{2^n+1} \cap H_{2^n+2} \cap \ldots \cap H_{2^n+[2\log_2 n]} \ i.o.\right\} = 0\,,$$

since in this case the probabilities are approximately $1/n^2$ which have finite sum.

An event like $\mathbf{P}(B_n \ i.o.)$, where $B_n = \{H_n \cap H_{n+1}\}$, is more difficult. In this case $\sum \mathbf{P}(B_n) = \sum_n(1/4) = \infty$. However, the $\{B_n\}$ are *not* independent, since B_n and B_{n+1} both involve the same event H_{n+1} (i.e., the $(n+1)^{\text{st}}$ coin). Hence, Theorem 3.4.2 does not immediately apply. On the other hand, by considering the subsequence $n = 2k$ of indices, we see that $\{B_{2k}\}_{k=1}^{\infty}$ are independent, and $\sum_{k=1}^{\infty} \mathbf{P}(B_{2k}) = \sum_{k=1}^{\infty} \mathbf{P}(B_{2k}) = \infty$. Hence, $\mathbf{P}(B_{2k} \ i.o.) = 1$, so that $\mathbf{P}(B_n \ i.o.) = 1$ also.

For a similar but more complicated example, let $B_n = \{H_{n+1} \cap H_{n+2} \cap \ldots \cap H_{n+[\log_2 \log_2 n]}\}$. Again, $\sum \mathbf{P}(B_n) = \infty$, but the $\{B_n\}$ are *not* independent. But by considering the subsequence $n = 2^k$ of indices, we compute that $\{B_{2k}\}$ are independent, and $\sum_k \mathbf{P}(B_{2k}) = \infty$. Hence, $\mathbf{P}(B_{2k} \ i.o.) = 1$, so that $\mathbf{P}(B_n \ i.o.) = 1$ also.

3.5. Tail fields.

Given a sequence of events $A_1, A_2, \ldots$, we define their *tail field* by

$$\tau = \bigcap_{n=1}^{\infty} \sigma(A_n, A_{n+1}, A_{n+2}, \ldots)\,.$$

In words, an event $A \in \tau$ must have the property that for any n, it depends only on the events $A_n, A_{n+1}, \ldots$; in particular, it does not care about any finite number of the events A_n.

One might think that very few events could possibly be in the tail field, but in fact it sometimes contains many events. For example, if we are considering infinite fair coin tossing (Subsection 2.6), and H_n is the event that the nth coin comes up heads, then τ includes the event $\limsup_n H_n$ that we obtain infinitely many heads; the event $\liminf_n H_n$ that we obtain only finitely many tails; the event $\limsup_n H_{2^n}$ that we obtain infinitely many heads on tosses $2, 4, 8, \ldots$; the event $\{\lim_{n \to \infty} \frac{1}{n} \sum_{i=1}^n r_i \le \frac{1}{4}\}$ that the limiting fraction of heads is $\le \frac{1}{4}$; the event $\{r_n = r_{n+1} = r_{n+2} \ i.o.\}$ that we infinitely often obtain the same result on three consecutive coin flips; etc. So we see that τ contains many interesting events.

A surprising theorem is

Theorem 3.5.1. *(Kolmogorov Zero-One Law.) If events $A_1, A_2, \ldots$ are independent, with tail-field τ, and if $A \in \tau$, then $\mathbf{P}(A) = 0$ or 1.*

To prove this theorem, we need a technical result about independence.

Lemma 3.5.2. *Let $B, B_1, B_2, \ldots$ be independent. Then $\{B\}$ and $\sigma(B_1, B_2, \ldots)$ are independent classes, i.e. if $S \in \sigma(B_1, B_2, \ldots)$, then $\mathbf{P}(S \cap B) = \mathbf{P}(S)\,\mathbf{P}(B)$.*

Proof. Assume that $\mathbf{P}(B) > 0$, otherwise the statement is trivial.

Let $\mathcal{J}$ be the collection of all sets of the form $D_{i_1} \cap D_{i_2} \cap \ldots \cap D_{i_n}$, where $n \in \mathbf{N}$ and where D_{i_j} is either B_{i_j} or $B_{i_j}^C$, together with $\emptyset$ and Ω. Then for $A \in \mathcal{J}$, we have by independence that $\mathbf{P}(A) = \mathbf{P}(B \cap A)/\mathbf{P}(B)$.

Now define a new probability measure $\mathbf{Q}$ on $\sigma(B_1, B_2, \ldots)$ by $\mathbf{Q}(S) = \mathbf{P}(B \cap S)/\mathbf{P}(B)$, for $S \in \sigma(B_1, B_2, \ldots)$. Then $\mathbf{Q}(\emptyset) = 0$, $\mathbf{Q}(\Omega) = 1$, and $\mathbf{Q}$ is countably additive since $\mathbf{P}$ is, so $\mathbf{Q}$ is indeed a probability measure. Furthermore, $\mathbf{Q}$ and $\mathbf{P}$ agree on $\mathcal{J}$. Hence, by Proposition 2.5.8, $\mathbf{Q}$ and $\mathbf{P}$ agree on $\sigma(\mathcal{J}) = \sigma(B_1, B_2, \ldots)$. That is, $\mathbf{P}(S) = \mathbf{Q}(S) = \mathbf{P}(B \cap S)/\mathbf{P}(B)$ for all $S \in \sigma(B_1, B_2, \ldots)$, as required. ∎

Applying this lemma twice, we obtain:

Corollary 3.5.3. *Let $A_1, A_2, \ldots, B_1, B_2, \ldots$ be independent. Then if $S_1 \in \sigma(A_1, A_2, \ldots)$, then $S_1, B_1, B_2, \ldots$ are independent. Furthermore, the σ-algebras $\sigma(A_1, A_2, \ldots)$ and $\sigma(B_1, B_2, \ldots)$ are independent classes, i.e. if $S_1 \in \sigma(A_1, A_2, \ldots)$, and $S_2 \in \sigma(B_1, B_2, \ldots)$, then $\mathbf{P}(S_1 \cap S_2) = \mathbf{P}(S_1)\,\mathbf{P}(S_2)$.*

Proof. For any distinct $i_1, i_2, \ldots, i_n$, let $A = B_{i_1} \cap \ldots \cap B_{i_n}$. Then it follows immediately that $A, A_1, A_2, \ldots$ are independent. Hence, from Lemma 3.5.2, if $S_1 \in \sigma(A_1, A_2, \ldots)$, then $\mathbf{P}(A \cap S_1) = \mathbf{P}(A)\,\mathbf{P}(S_1)$. Since this is true for all distinct $i_1, \ldots, i_n$, it follows that $S_1, B_1, B_2, \ldots$ are independent. Lemma 3.5.2 then implies that if $S_2 \in \sigma(B_1, B_2, \ldots)$, then S_1 and

S_2 are independent.   ∎

Proof of Theorem 3.5.1.   We can now easily prove the Kolmogorov Zero-One Law. The proof is rather remarkable!

Indeed, $A \in \sigma(A_n, A_{n+1}, \ldots)$, so by Corollary 3.5.3, $A, A_1, A_2, \ldots, A_{n-1}$ are independent. Since this is true for all $n \in \mathbf{N}$, and since independence is defined in terms of finite subcollections only, it is also true that $A, A_1, A_2, \ldots$ are independent. Hence, from Lemma 3.5.2, A and S are independent for all $S \in \sigma(A_1, A_2, \ldots)$.

On the other hand, $A \in \tau \subseteq \sigma(A_1, A_2, \ldots)$. It follows that A is independent of itself (!). This implies that $\mathbf{P}(A \cap A) = \mathbf{P}(A)\,\mathbf{P}(A)$. That is, $\mathbf{P}(A) = \mathbf{P}(A)^2$, so $\mathbf{P}(A) = 0$ or 1.   ∎

3.6. Exercises.

Exercise 3.6.1.   Let X be a real-valued random variable defined on a probability triple $(\Omega, \mathcal{F}, \mathbf{P})$. Fill in the following blanks:
(a) $\mathcal{F}$ is a collection of subsets of _____.
(b) $\mathbf{P}(A)$ is a well-defined element of _____ provided that A is an element of _____.
(c) $\{X \leq 5\}$ is shorthand notation for the particular subset of _____ which is defined by: _____.
(d) If S is a subset of _____, then $\{X \in S\}$ is a subset of _____.
(e) If S is a _____ subset of _____, then $\{X \in S\}$ must be an element of _____.

Exercise 3.6.2.   Let $(\Omega, \mathcal{F}, \mathbf{P})$ be Lebesgue measure on $[0, 1]$. Let $A = (1/2, 3/4)$ and $B = (0, 2/3)$. Are A and B independent events?

Exercise 3.6.3.   Give an example of events A, B, and C, each of probability strictly between 0 and 1, such that
(a) $\mathbf{P}(A \cap B) = \mathbf{P}(A)\mathbf{P}(B)$, $\mathbf{P}(A \cap C) = \mathbf{P}(A)\mathbf{P}(C)$, and $\mathbf{P}(B \cap C) = \mathbf{P}(B)\mathbf{P}(C)$; but it is *not* the case that $\mathbf{P}(A \cap B \cap C) = \mathbf{P}(A)\mathbf{P}(B)\mathbf{P}(C)$. [Hint: You can let Ω be a set of four equally likely points.]
(b) $\mathbf{P}(A \cap B) = \mathbf{P}(A)\mathbf{P}(B)$, $\mathbf{P}(A \cap C) = \mathbf{P}(A)\mathbf{P}(C)$, and $\mathbf{P}(A \cap B \cap C) = \mathbf{P}(A)\mathbf{P}(B)\mathbf{P}(C)$; but it is *not* the case that $\mathbf{P}(B \cap C) = \mathbf{P}(B)\mathbf{P}(C)$. [Hint: You can let Ω be a set of eight equally likely points.]

Exercise 3.6.4.   Suppose $\{A_n\} \nearrow A$. Let $f : \Omega \to \mathbf{R}$ be any function. Prove that $\lim_{n \to \infty} \inf_{\omega \in A_n} f(\omega) = \inf_{\omega \in A} f(\omega)$.

Exercise 3.6.5. Let $(\Omega, \mathcal{F}, \mathbf{P})$ be a probability triple such that Ω is *countable*. Prove that it is impossible for there to exist a sequence $A_1, A_2, \ldots \in \mathcal{F}$ which is *independent*, such that $\mathbf{P}(A_i) = \frac{1}{2}$ for each i. [**Hint:** First prove that for each $\omega \in \Omega$, and each $n \in \mathbf{N}$, we have $P(\{\omega\}) \leq 1/2^n$. Then derive a contradiction.]

Exercise 3.6.6. Let X, Y, and Z be three independent random variables, and set $W = X + Y$. Let $B_{k,n} = \{(n-1)2^{-k} \leq X < n2^{-k}\}$ and let $C_{k,m} = \{(m-1)2^{-k} \leq Y < m2^{-k}\}$. Let

$$
A_k = \overset{\bullet}{\underset{\substack{n,m \in \mathbf{Z} \\ (n+m)2^{-k} < x}}{\bigcup}} (B_{k,n} \cap C_{k,m}).
$$

Fix $x, z \in \mathbf{R}$, and let $A = \{X + Y < x\} = \{W < x\}$ and $D = \{Z < z\}$.
(a) Prove that $\{A_k\} \nearrow A$.
(b) Prove that A_k and D are independent.
(c) By continuity of probabilities, prove that A and D are independent.
(d) Use this to prove that W and Z are independent.

Exercise 3.6.7. Let $(\Omega, \mathcal{F}, \mathbf{P})$ be the uniform distribution on $\Omega = \{1, 2, 3\}$, as in Example 2.2.2. Give an example of a sequence $A_1, A_2, \ldots \in \mathcal{F}$ such that

$$
\mathbf{P}\left(\liminf_n A_n\right) < \liminf_n \mathbf{P}(A_n) < \limsup_n \mathbf{P}(A_n) < \mathbf{P}\left(\limsup_n A_n\right),
$$

i.e. such that all three inequalities are *strict*.

Exercise 3.6.8. Let λ be Lebesgue measure on $[0, 1]$, and let $0 \leq a \leq b \leq c \leq d \leq 1$ be arbitrary real numbers. Give an example of a sequence $A_1, A_2, \ldots$ of subsets of $[0, 1]$, such that $\lambda(\liminf_n A_n) = a$, $\liminf_n \lambda(A_n) = b$, $\limsup_n \lambda(A_n) = c$, and $\lambda(\limsup_n A_n) = d$. [Hint: begin with the case $d = b + c - a$, which is easiest, and then carefully branch out from there.]

Exercise 3.6.9. Let $A_1, A_2, \ldots, B_1, B_2, \ldots$ be events.
(a) Prove that

$$
\left(\limsup_n A_n\right) \cap \left(\limsup_n B_n\right) \supseteq \limsup_n (A_n \cap B_n).
$$

(b) Give an example where the above inclusion is strict, and another example where it holds with equality.

Exercise 3.6.10.   Let $A_1, A_2, \ldots$ be a sequence of events, and let $N \in \mathbf{N}$. Suppose there are events B and C such that $B \subseteq A_n \subseteq C$ for all $n \geq N$, and such that $\mathbf{P}(B) = \mathbf{P}(C)$. Prove that $\mathbf{P}(\liminf_n A_n) = \mathbf{P}(\limsup_n A_n) = \mathbf{P}(B) = \mathbf{P}(C)$.

Exercise 3.6.11.   Let $\{X_n\}_{n=1}^\infty$ be independent random variables, with $X_n \sim \mathbf{Uniform}(\{1, 2, \ldots, n\})$ (cf. Example 2.2.2). Compute $\mathbf{P}(X_n = 5 \; i.o.)$, the probability that an infinite number of the X_n are equal to 5.

Exercise 3.6.12.   Let X be a random variable with $\mathbf{P}(X > 0) > 0$. Prove that there is $\delta > 0$ such that $\mathbf{P}(X \geq \delta) > 0$. [Hint: Don't forget continuity of probabilities.]

Exercise 3.6.13.   Let $X_1, X_2, \ldots$ be defined jointly on some probability space $(\Omega, \mathcal{F}, P)$, with $E[X_i] = 0$ and $E[(X_i)^2] = 1$ for all i. Prove that $\mathbf{P}[X_n \geq n \; i.o.] = 0$.

Exercise 3.6.14.   Let $\delta, \epsilon > 0$, and let $X_1, X_2, \ldots$ be a sequence of independent non-negative random variables such that $\mathbf{P}(X_i \geq \delta) \geq \epsilon$ for all i. Prove that with probability one, $\sum_{i=1}^\infty X_i = \infty$.

Exercise 3.6.15.   Let $A_1, A_2, \ldots$ be a sequence of events, such that (i) $A_{i_1}, A_{i_2}, \ldots, A_{i_k}$ are independent whenever $i_{j+1} \geq i_j + 2$ for $1 \leq j \leq k - 1$, and (ii) $\sum_n \mathbf{P}(A_n) = \infty$. Then the Borel-Cantelli Lemma does not directly apply. Still, prove that $\mathbf{P}(\limsup_n A_n) = 1$.

Exercise 3.6.16.   Consider infinite, independent, fair coin tossing as in Subsection 2.6, and let H_n be the event that the nth coin is heads. Determine the following probabilities.
(a) $\mathbf{P}(H_{n+1} \cap H_{n+2} \cap \ldots \cap H_{n+9} \; i.o.)$.
(b) $\mathbf{P}(H_{n+1} \cap H_{n+2} \cap \ldots \cap H_{2n} \; i.o.)$.
(c) $\mathbf{P}(H_{n+1} \cap H_{n+2} \cap \ldots \cap H_{n+[2\log_2 n]} \; i.o.)$.
(d) Prove that $\mathbf{P}(H_{n+1} \cap H_{n+2} \cap \ldots \cap H_{n+[\log_2 n]} \; i.o.)$ must equal either 0 or 1.
(e) Determine $\mathbf{P}(H_{n+1} \cap H_{n+2} \cap \ldots \cap H_{n+[\log_2 n]} \; i.o.)$. [Hint: Find the right subsequence of indices.]

Exercise 3.6.17.   Show that Lemma 3.5.2 is false if we require only that $\mathbf{P}(B \cap B_n) = \mathbf{P}(B)\,\mathbf{P}(B_n)$ for each $n \in \mathbf{N}$, but do not require that the $\{B_n\}$ be independent of each other. [Hint: Don't forget Exercise 3.6.3(a).]

Exercise 3.6.18.   Let $A_1, A_2, \ldots$ be any independent sequence of events, and let $S_x = \{\lim_{n\to\infty} \frac{1}{n} \sum_{i=1}^n \mathbf{1}_{A_i} \leq x\}$. Prove that for each $x \in \mathbf{R}$ we have $\mathbf{P}(S_x) = 0$ or 1.

Exercise 3.6.19. Let $A_1, A_2, \ldots$ be independent events. Let Y be a random variable which is measurable with respect to $\sigma(A_n, A_{n+1}, \ldots)$ for each $n \in \mathbf{N}$. Prove that there is a real number a such that $\mathbf{P}(Y = a) = 1$. [Hint: Consider $\mathbf{P}(Y \leq x)$ for $x \in \mathbf{R}$; what values can it take?]

3.7. Section summary.

In this section, we defined random variables, which are functions on the state space. We also defined independence of events and of random variables. We derived the continuity property of probability measures. We defined limit events and proved the important Borel-Cantelli Lemma. We defined tail fields and proved the remarkable Kolmogorov Zero-One Law.

4. Expected values.

There is one more notion that is fundamental to all of probability theory, that of expected values. The general definition of expected value will be developed in this section.

4.1. Simple random variables.

Let $(\Omega, \mathcal{F}, \mathbf{P})$ be a probability triple, and let X be a random variable defined on this triple. We begin with a definition.

Definition 4.1.1. A random variable X is *simple* if range(X) is finite, where range(X) $\equiv \{X(\omega); \omega \in \Omega\}$.

That is, a random variable is simple if it takes on only a finite number of different values. If X is a simple random variable, then listing the distinct elements of its range as $x_1, x_2, \ldots, x_n$, we can then write $X = \sum_{i=1}^{n} x_i \mathbf{1}_{A_i}$ where $A_i = \{\omega \in \Omega; X(\omega) = x_i\} = X^{-1}(\{x_i\})$, and where the $\mathbf{1}_{A_i}$ are indicator functions. We note that the sets A_i form a finite partition of Ω.

For such a simple random variable $X = \sum_{i=1}^{n} x_i \mathbf{1}_{A_i}$, we define its *expected value* or *expectation* or *mean* by $\mathbf{E}(X) = \sum_{i=1}^{n} x_i \mathbf{P}(A_i)$. That is,

$$\mathbf{E}\left(\sum_{i=1}^{n} x_i \mathbf{1}_{A_i}\right) = \sum_{i=1}^{n} x_i \mathbf{P}(A_i), \quad \{A_i\} \text{ a finite partition of } \Omega. \quad (4.1.2)$$

We sometimes write μ_X for $\mathbf{E}(X)$.

Exercise 4.1.3. Prove that (4.1.2) is well-defined, in the sense that if $\{A_i\}$ and $\{B_j\}$ are two different finite partitions of Ω, such that $\sum_{i=1}^{n} x_i \mathbf{1}_{A_i} = \sum_{j=1}^{m} y_j \mathbf{1}_{B_j}$, then $\sum_{i=1}^{n} x_i \mathbf{P}(A_i) = \sum_{j=1}^{m} y_j \mathbf{P}(B_j)$. [Hint: collect together those A_i and B_j corresponding to the same values of x_i and y_j.]

For a quick example, let $(\Omega, \mathcal{F}, \mathbf{P})$ be Lebesgue measure on $[0, 1]$, and define simple random variables X and Y by

$$X(\omega) = \begin{cases} 5, & \omega > 1/3 \\ 3, & \omega \le 1/3, \end{cases} \qquad Y(\omega) = \begin{cases} 2, & \omega \text{ rational} \\ 4, & \omega = 1/\sqrt{2} \\ 6, & \text{other } \omega \le 1/4 \\ 8, & \text{otherwise}. \end{cases}$$

Then it is easily seen that $\mathbf{E}(X) = 13/3$, and $\mathbf{E}(Y) = 15/2$.

From equation (4.1.2), we see immediately that $\mathbf{E}(\mathbf{1}_A) = \mathbf{P}(A)$, and that $\mathbf{E}(c) = c$. We now claim that $\mathbf{E}(\cdot)$ is linear. Indeed, if $X = \sum_i x_i \mathbf{1}_{A_i}$

and $Y = \sum_j y_j \mathbf{1}_{B_j}$, where $\{A_i\}$ and $\{B_j\}$ are finite partitions of Ω, and if $a, b, \in \mathbf{R}$, then $\{A_i \cap B_j\}$ is again a finite partition of Ω, and we have

$$
\begin{aligned}
\mathbf{E}(aX + bY) &= \mathbf{E}\left(\sum_{i,j}(ax_i + by_j)\mathbf{1}_{A_i \cap B_j}\right) \\
&= \sum_{i,j}(ax_i + by_j)\mathbf{P}(A_i \cap B_j) \\
&= a\sum_i x_i \mathbf{P}(A_i) + b\sum_j y_j \mathbf{P}(B_j) \\
&= a\mathbf{E}(X) + b\mathbf{E}(Y),
\end{aligned}
$$

as claimed. It follows that $\mathbf{E}\left(\sum_{i=1}^n x_i \mathbf{1}_{A_i}\right) = \sum_{i=1}^n x_i \mathbf{P}(A_i)$ for *any* finite collection of subsets $A_i \subseteq \Omega$, even if they do not form a partition.

It also follows that $\mathbf{E}(\cdot)$ is *order-preserving*, i.e. if $X \leq Y$ (meaning that $X(\omega) \leq Y(\omega)$ for all $\omega \in \Omega$), then $\mathbf{E}(X) \leq \mathbf{E}(Y)$. Indeed, in that case $Y - X \geq 0$, so from (4.1.2) we have $\mathbf{E}(Y - X) \geq 0$; by linearity this implies that $\mathbf{E}(Y) - \mathbf{E}(X) \geq 0$.

In particular, since $-|X| \leq X \leq |X|$, we have $|\mathbf{E}(X)| \leq \mathbf{E}(|X|)$, which is sometimes referred to as the *(generalised) triangle inequality*. If X takes on the two values a and b, each with probability $\frac{1}{2}$, then this inequality reduces to the usual $|a + b| \leq |a| + |b|$.

Finally, if X and Y are *independent* simple random variables, then $\mathbf{E}(XY) = \mathbf{E}(X)\mathbf{E}(Y)$. Indeed, again writing $X = \sum_i x_i \mathbf{1}_{A_i}$ and $Y = \sum_j y_j \mathbf{1}_{B_j}$, where $\{A_i\}$ and $\{B_j\}$ are finite partitions of Ω and where $\{x_i\}$ are distinct and $\{y_j\}$ are distinct, we see that X and Y are independent if and only if $\mathbf{P}(A_i \cap B_j) = \mathbf{P}(A_i)\mathbf{P}(B_j)$ for all i and j. In that case, $\mathbf{E}(XY) = \sum_{i,j} x_i y_j \mathbf{P}(A_i \cap B_j) = \sum_{i,j} x_i y_j \mathbf{P}(A_i)\mathbf{P}(B_j) = \mathbf{E}(X)\mathbf{E}(Y)$, as claimed. Note that this may be false if X and Y are not independent; for example, if X takes on the values ± 1, each with probability $\frac{1}{2}$, and if $Y = X$, then $\mathbf{E}(X) = \mathbf{E}(Y) = 0$ but $\mathbf{E}(XY) = 1$. Also, we may have $\mathbf{E}(XY) = \mathbf{E}(X)\mathbf{E}(Y)$ even if X and Y are *not* independent; for example, this occurs if X takes on the three values 0, 1, and 2 each with probability $\frac{1}{3}$, and if Y is defined by $Y(\omega) = 1$ whenever $X(\omega) = 0$ or 2, and $Y(\omega) = 5$ whenever $X(\omega) = 1$.

If $X = \sum_{i=1}^n x_i \mathbf{1}_{A_i}$, with $\{A_i\}$ a finite partition of Ω, and if $f : \mathbf{R} \to \mathbf{R}$ is any function, then $f(X) = \sum_{i=1}^n f(x_i)\mathbf{1}_{A_i}$ is also a simple random variable, with $\mathbf{E}(f(X)) = \sum_{i=1}^n f(x_i)\mathbf{P}(A_i)$.

In particular, if $f(x) = (x - \mu_X)^2$, we get the *variance* of X, defined by $\mathbf{Var}(X) = \mathbf{E}\left((X - \mu_X)^2\right)$. Clearly $\mathbf{Var}(X) \geq 0$. Expanding the square and using linearity, we see that $\mathbf{Var}(X) = \mathbf{E}(X^2) - \mu_X^2 = \mathbf{E}(X^2) - \mathbf{E}(X)^2$. In particular, we always have

$$
\mathbf{Var}(X) \leq \mathbf{E}(X^2). \tag{4.1.4}
$$

It also follows immediately that

$$
\mathbf{Var}(\alpha X + \beta) = \alpha^2 \mathbf{Var}(X). \tag{4.1.5}
$$

We also see that $\mathbf{Var}(X + Y) = \mathbf{Var}(X) + \mathbf{Var}(Y) + 2\,\mathbf{Cov}(X, Y)$, where $\mathbf{Cov}(X, Y) = \mathbf{E}\left((X - \mu_X)(Y - \mu_Y)\right) = \mathbf{E}(XY) - \mathbf{E}(X)\mathbf{E}(Y)$ is the covariance; in particular, if X and Y are independent then $\mathbf{Cov}(X, Y) = 0$ so that $\mathbf{Var}(X + Y) = \mathbf{Var}(X) + \mathbf{Var}(Y)$. More generally, $\mathbf{Var}(\sum_i X_i) = \sum_i \mathbf{Var}(X_i) + 2\sum_{i<j} \mathbf{Cov}(X_i, X_j)$, so we see that

$$\mathbf{Var}(X_1 + \ldots + X_n) = \mathbf{Var}(X_1) + \ldots + \mathbf{Var}(X_n), \quad \{X_n\} \text{ independent}.$$
(4.1.6)

Finally, if $\mathbf{Var}(X) > 0$ and $\mathbf{Var}(Y) > 0$, then the *correlation* between X and Y is defined by $\mathbf{Corr}(X, Y) = \mathbf{Cov}(X, Y)/\sqrt{\mathbf{Var}(X)\,\mathbf{Var}(Y)}$; see Exercises 4.5.11 and 5.5.6.

This concludes our discussion of the basic properties of expectation for simple random variables. (Indeed, it is possible to read Section 5 immediately at this point, provided that one restricts attention to simple random variables only.) We now note a fact that will help us to define $\mathbf{E}(X)$ for random variables X which are *not* simple. It follows immediately from the order-preserving property of $\mathbf{E}(\cdot)$.

Proposition 4.1.7. *If X is a simple random variable, then*

$$\mathbf{E}(X) = \sup\{\mathbf{E}(Y)\,;\, Y \text{ simple}, Y \leq X\}.$$

4.2. General non-negative random variables.

If X is not simple, then it is not clear how to define its expected value $\mathbf{E}(X)$. However, Proposition 4.1.7 provides a suggestion of how to proceed. Indeed, for a general non-negative random variable X, we define the expected value $\mathbf{E}(X)$ by

$$\mathbf{E}(X) = \sup\{\mathbf{E}(Y)\,;\, Y \text{ simple}, Y \leq X\}.$$

By Proposition 4.1.7, if X happens to be a simple random variable then this definition agrees with the previous one, so there is no confusion in reusing the same symbol $\mathbf{E}(\cdot)$. Indeed, this one single definition (with a minor modification for negative values in Subsection 4.3 below) will apply to all random variables, be they discrete, absolutely continuous, or neither (cf. Section 6).

We note that it is indeed possible that $\mathbf{E}(X)$ will be *infinite*. For example, suppose $(\Omega, \mathcal{F}, \mathbf{P})$ is Lebesgue measure on $[0, 1]$, and define X by $X(\omega) = 2^n$ for $2^{-n} \leq \omega < 2^{-(n-1)}$. (See Figure 4.2.1.) Then $\mathbf{E}(X) \geq \sum_{k=1}^N 2^k 2^{-k} = N$ for any $N \in \mathbf{N}$. Hence, $\mathbf{E}(X) = \infty$.

Recall that for $k \in \mathbf{N}$, the kth *moment* of a non-negative random variable X is defined to be $\mathbf{E}(X^k)$, finite if $\mathbf{E}(|X|^k) < \infty$. Since $|x|^{k-1} \leq$

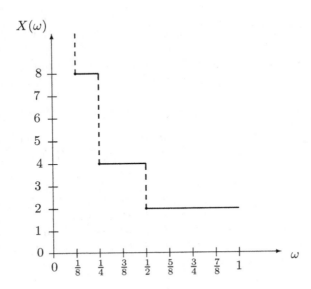

Figure 4.2.1. A random variable X having $\mathbf{E}(X) = \infty$.

$\max(|x|^k, 1) \leq |x|^k + 1$ for any $x \in \mathbf{R}$, we see that if $\mathbf{E}(|X|^k)$ is finite, then so is $\mathbf{E}(|X|^{k-1})$.

It is immediately apparent that our general definition of $\mathbf{E}(\cdot)$ is still order-preserving. However, proving linearity is less clear. To assist, we have the following result. We say that $\{X_n\} \nearrow X$ if $X_1 \leq X_2 \leq \ldots$, and also $\lim_{n \to \infty} X_n(\omega) = X(\omega)$ for each $\omega \in \Omega$. (That is, the sequence $\{X_n\}$ converges *monotonically* to X.)

Theorem 4.2.2.   (*The monotone convergence theorem.*) *Suppose* $X_1, X_2, \ldots$ *are random variables with* $\mathbf{E}(X_1) > -\infty$, *and* $\{X_n\} \nearrow X$. *Then* X *is a random variable, and* $\lim_{n \to \infty} \mathbf{E}(X_n) = \mathbf{E}(X)$.

Proof.   We know from (3.1.6) that X is a random variable (alternatively, simply note that in this case, $\{X \leq x\} = \bigcap_n \{X_n \leq x\} \in \mathcal{F}$ for all $x \in \mathbf{R}$). Furthermore, by monotonicity we have $\mathbf{E}(X_1) \leq \mathbf{E}(X_2) \leq \ldots \leq \mathbf{E}(X)$, so that $\lim_n \mathbf{E}(X_n)$ exists (though it may be infinite if $\mathbf{E}(X) = +\infty$), and is $\leq \mathbf{E}(X)$.

To finish, it suffices to show that $\lim_n \mathbf{E}(X_n) \geq \mathbf{E}(X)$. If $E(X_1) = +\infty$ this is trivial, so assume $E(X_1)$ is finite. Then, by replacing X_n by $X_n - X_1$ and X by $X - X_1$, it suffices to assume the X_n and X are non-negative.

By the definition of $\mathbf{E}(X)$ for non-negative X, it suffices to show that $\lim_n \mathbf{E}(X_n) \geq \mathbf{E}(Y)$ for any simple random variable $Y \leq X$. Writing $Y = \sum_i v_i \mathbf{1}_{A_i}$, we see that it suffices to prove that $\lim_n \mathbf{E}(X_n) \geq \sum_i v_i \mathbf{P}(A_i)$, where $\{A_i\}$ is any finite partition of Ω with $v_i \leq X(\omega)$ for all $\omega \in A_i$.

To that end, choose $\epsilon > 0$, and set $A_{in} = \{\omega \in A_i \,;\, X_n(\omega) \geq v_i - \epsilon\}$. Then $\{A_{in}\} \nearrow A_i$ as $n \to \infty$. Furthermore, $\mathbf{E}(X_n) \geq \sum_i (v_i - \epsilon) \mathbf{P}(A_{in})$. As $n \to \infty$, by continuity of probabilities this converges to $\sum_i (v_i - \epsilon) \mathbf{P}(A_i) = \sum_i v_i \mathbf{P}(A_i) - \epsilon$. Hence, $\lim \mathbf{E}(X_n) \geq \sum_i v_i \mathbf{P}(A_i) - \epsilon$. Since this is true for any $\epsilon > 0$, we must (cf. Proposition A.3.1) have $\lim \mathbf{E}(X_n) \geq \sum_i v_i \mathbf{P}(A_i)$, as required. ∎

Remark 4.2.3. Since expected values are unchanged if we modify the random variable values on sets of probability 0, we still have $\lim_{n \to \infty} \mathbf{E}(X_n) = \mathbf{E}(X)$ provided $\{X_n\} \nearrow X$ *almost surely (a.s.)*, i.e. on a subset of Ω having probability 1. (Compare Remark 3.1.9.)

We note that the monotonicity assumption of Theorem 4.2.2 is indeed necessary. For example, if $(\Omega, \mathcal{F}, \mathbf{P})$ is Lebesgue measure on $[0, 1]$, and if $X_n = n \mathbf{1}_{(0, \frac{1}{n})}$, then $X_n \to 0$ (since for each $\omega \in [0, 1]$ we have $X_n(\omega) = 0$ for all $n \geq 1/\omega$), but $\mathbf{E}(X_n) = 1$ for all n.

To make use of Theorem 4.2.2, set $\Psi_n(x) = \min(n, 2^{-n} \lfloor 2^n x \rfloor)$ for $x \geq 0$, where $\lfloor r \rfloor$ is the *floor* of r, or greatest integer not exceeding r. (See Figure 4.2.4.) Then $\Psi_n(x)$ is a slightly rounded-down version of x, truncated at n. Indeed, for fixed $x \geq 0$ we have that $\Psi(x) \geq 0$, and $\{\Psi_n(x)\} \nearrow x$ as $n \to \infty$. Furthermore, the range of Ψ_n is finite (of size $n2^n + 1$). Hence, this shows:

Proposition 4.2.5. *Let X be a general non-negative random variable. Set $X_n = \Psi_n(X)$ with Ψ_n as above. Then $X_n \geq 0$ and $\{X_n\} \nearrow X$, and each X_n is a simple random variable. In particular, there exists a sequence of simple random variables increasing to X.*

Using Theorem 4.2.2 and Proposition 4.2.5, it is straightforward to prove the linearity of expected values for general non-negative random variables. Indeed, with $X_n = \Psi_n(X)$ and $Y_n = \Psi_n(Y)$, we have (using linearity of expectation for simple random variables) that for $X, Y \geq 0$ and $a, b \geq 0$,

$$\mathbf{E}(aX + bY) = \lim_n \mathbf{E}(aX_n + bY_n) = \lim_n \left(a\mathbf{E}(X_n) + b\mathbf{E}(Y_n) \right)$$

$$= a\mathbf{E}(X) + b\mathbf{E}(Y) . \tag{4.2.6}$$

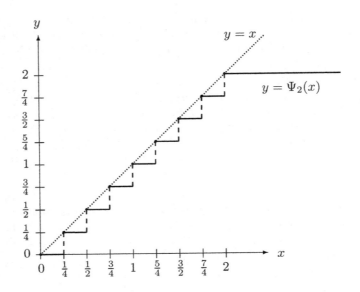

Figure 4.2.4. A graph of $y = \Psi_2(x)$.

Similarly, if X and Y are independent, then by Proposition 3.2.3, X_n and Y_n are also independent. Hence, if $\mathbf{E}(X)$ and $\mathbf{E}(Y)$ are finite,

$$\mathbf{E}(XY) = \lim_n \mathbf{E}(X_n Y_n) = \lim_n \mathbf{E}(X_n)\,\mathbf{E}(Y_n) = \mathbf{E}(X)\,\mathbf{E}(Y)\,, \qquad (4.2.7)$$

as was the case for simple random variables. It then follows that $\mathbf{Var}(X + Y) = \mathbf{Var}(X) + \mathbf{Var}(Y)$ exactly as in the previous case.

We also have *countable linearity*. Indeed, if $X_1, X_2, \ldots \geq 0$, then by the monotone convergence theorem,

$$\mathbf{E}(X_1 + X_2 + \ldots) = \mathbf{E}\left(\lim_n X_1 + X_2 + \ldots + X_n\right) = \lim_n \mathbf{E}(X_1 + X_2 + \ldots + X_n)$$

$$= \lim_n \left[\mathbf{E}(X_1) + \mathbf{E}(X_2) + \ldots + \mathbf{E}(X_n)\right] = \mathbf{E}(X_1) + \mathbf{E}(X_2) + \ldots\,. \qquad (4.2.8)$$

If the X_i are not non-negative, then (4.2.8) may fail, though it still holds under certain conditions; see Exercise 4.5.14 (and Corollary 9.4.4).

We also have the following.

Proposition 4.2.9.   *If X is a non-negative random variable, then* $\sum_{k=1}^{\infty} \mathbf{P}(X \geq k) = \mathbf{E}\lfloor X \rfloor$, *where $\lfloor X \rfloor$ is the greatest integer not exceeding*

X. *(In particular, if X is non-negative-integer valued, then $\sum_{k=1}^{\infty} \mathbf{P}(X \geq k) = \mathbf{E}(X)$.)*

Proof. We compute that

$$\sum_{k=1}^{\infty} \mathbf{P}(X \geq k) = \sum_{k=1}^{\infty} [\mathbf{P}(k \leq X < k+1) + \mathbf{P}(k+1 \leq X < k+2) + \ldots]$$

$$= \sum_{\ell=1}^{\infty} \ell\, \mathbf{P}(\ell \leq X < \ell+1) = \sum_{\ell=1}^{\infty} \ell\, \mathbf{P}(\lfloor X \rfloor = \ell) = \mathbf{E}\lfloor X \rfloor. \quad \blacksquare$$

4.3. Arbitrary random variables.

Finally, we consider random variables which may be neither simple nor non-negative. For such a random variable X, we may write $X = X^+ - X^-$, where $X^+(\omega) = \max(X(\omega), 0)$ and $X^-(\omega) = \max(-X(\omega), 0)$. Both X^+ and X^- are non-negative, so the theory of the previous subsection applies to them. We may then set

$$\mathbf{E}(X) = \mathbf{E}(X^+) - \mathbf{E}(X^-). \qquad (4.3.1)$$

We note that $\mathbf{E}(X)$ is *undefined* if both $\mathbf{E}(X^+)$ and $\mathbf{E}(X^-)$ are infinite. However, if $\mathbf{E}(X^+) = \infty$ and $\mathbf{E}(X^-) < \infty$, then we take $\mathbf{E}(X) = \infty$. Similarly, if $\mathbf{E}(X^+) < \infty$ and $\mathbf{E}(X^-) = \infty$, then we take $\mathbf{E}(X) = -\infty$. Obviously, if $\mathbf{E}(X^+) < \infty$ and $\mathbf{E}(X^-) < \infty$, then $\mathbf{E}(X)$ will be a finite number.

We next check that, with this modification, expected value retains the basic properties of order-preserving, linear, etc.:

Exercise 4.3.2. Let X and Y be two general random variables (not necessarily non-negative) with well-defined means, such that $X \leq Y$.
(a) Prove that $X^+ \leq Y^+$ and $X^- \geq Y^-$.
(b) Prove that expectation is still order-preserving, i.e. that $\mathbf{E}(X) \leq \mathbf{E}(Y)$ under these assumptions.

Exercise 4.3.3. Let X and Y be two general random variables with finite means, and let $Z = X + Y$.
(a) Express Z^+ and Z^- in terms of X^+, X^-, Y^+, and Y^-.
(b) Prove that $\mathbf{E}(Z) = \mathbf{E}(X) + \mathbf{E}(Y)$, i.e. that $\mathbf{E}(Z^+) - \mathbf{E}(Z^-) = \mathbf{E}(X^+) - \mathbf{E}(X^-) + \mathbf{E}(Y^+) - \mathbf{E}(Y^-)$. [Hint: Re-arrange the relations of part (a) so that you can make use of (4.2.6).]

(c) Prove that expectation is still (finitely) linear, for general random variables with finite means.

Exercise 4.3.4.   Let X and Y be two independent general random variables with finite means, and let $Z = XY$.
(a) Prove that X^+ and Y^+ are independent, and similarly for each of X^+ and Y^-, and X^- and Y^+, and X^- and Y^-.
(b) Express Z^+ and Z^- in terms of X^+, X^-, Y^+, and Y^-.
(c) Prove that $\mathbf{E}(XY) = \mathbf{E}(X)\,\mathbf{E}(Y)$.

4.4. The integration connection.

Given a probability triple $(\Omega, \mathcal{F}, \mathbf{P})$, we sometimes write $\mathbf{E}(X)$ as $\int_\Omega X\, d\mathbf{P}$ or $\int_\Omega X(\omega)\mathbf{P}(d\omega)$ (the "Ω" is sometimes omitted). We call this *the integral of the (measurable) function X with respect to the (probability) measure* $\mathbf{P}$.

Why do we make this identification? Well, certainly expected value satisfies some similar properties to that of the integral: it is linear, order-preserving, etc. But a more convincing reason is given by

Theorem 4.4.1.   Let $(\Omega, \mathcal{F}, \mathbf{P})$ be Lebesgue measure on $[0,1]$. Let $X : [0,1] \to \mathbf{R}$ be a bounded function which is Riemann integrable (i.e. integrable in the usual calculus sense). Then X is a random variable with respect to $(\Omega, \mathcal{F}, \mathbf{P})$, and $\mathbf{E}(X) = \int_0^1 X(t)\, dt$. In words, the expected value of the random variable X is equal to the calculus-style integral of the function X.

Proof.   Recall the definitions of *lower* and *upper integrals*, viz.

$$L\int_0^1 X \equiv \sup\left\{ \sum_{i=1}^n (t_i - t_{i-1}) \inf_{t_{i-1}\leq t\leq t_i} X(t)\,;\, 0 = t_0 < t_1 < \ldots < t_n = 1 \right\}\,;$$

$$U\int_0^1 X \equiv \inf\left\{ \sum_{i=1}^n (t_i - t_{i-1}) \sup_{t_{i-1}\leq t\leq t_i} X(t)\,;\, 0 = t_0 < t_1 < \ldots < t_n = 1 \right\}\,.$$

Recall that we always have $L\int_0^1 X \leq U\int_0^1 X$, and that X is *Riemann integrable* if $L\int_0^1 X = U\int_0^1 X$, in which case $\int_0^1 X(t)\, dt$ is defined to be this common value.

But $\sum_{i=1}^n (t_i - t_{i-1}) \inf_{t_{i-1}\leq t\leq t_i} X(t) = \mathbf{E}(Y)$, where the simple random variable Y is defined by $Y(\omega) = \inf_{t_{i-1}\leq t\leq t_i} X(t)$ whenever $t_{i-1} \leq \omega < t_i$. Hence, since X is Riemann integrable, for each $n \in \mathbf{N}$ we can find a simple random variable $Y_n \leq X$ with $\mathbf{E}(Y_n) \geq L \int_0^1 X(t) \, dt - \frac{1}{n}$. Similarly we can find $Z_n \geq X$ with $\mathbf{E}(Z_n) \leq U \int_0^1 X(t) \, dt + \frac{1}{n}$. Let $A_n = \max(Y_1, \ldots, Y_n)$, and $B_n = \min(Z_1, \ldots, Z_n)$.

We claim that $\{A_n\} \nearrow X$ a.s. Indeed, $\{A_n\}$ is increasing by construction. Furthermore, $A_n \leq X \leq B_n$, and $\lim_{n \to \infty} \mathbf{E}(A_n) = \lim_{n \to \infty} \mathbf{E}(B_n) = \int_0^1 X(t) \, dt$. Hence, if $S_k = \{\omega \in \Omega : \lim_{n \to \infty} (B_n(\omega) - A_n(\omega)) \geq 1/k\}$, then $0 = \lim_{n \to \infty} \mathbf{E}(B_n - A_n) \geq \mathbf{P}(S_k) / k$, so $\mathbf{P}(S_k) = 0$ for all $k \in \mathbf{N}$. Then $\mathbf{P}\left[\lim_{n \to \infty} A_n < X\right] \leq \mathbf{P}\left[\lim_{n \to \infty} A_n < \lim_{n \to \infty} B_n\right] \leq \mathbf{P}\left[\bigcup_k S_k\right] = 0$ by countable subadditivity, proving the claim.

Hence, by Theorem 4.2.2 and Remarks 4.2.3 and 3.1.9, X must be a random variable, with $\mathbf{E}(X) = \lim_{n \to \infty} \mathbf{E}(A_n) = \int_0^1 X(t) \, dt$. ∎

On the other hand, there are many functions X which are *not* Riemann integrable, but which nevertheless are random variables with respect to Lebesgue measure, and thus have well-defined expected values. For example, if X is defined by $X(t) = 1$ for t irrational, but $X(t) = 0$ for t rational, i.e. $X = \mathbf{1}_{\mathbf{Q}^C}$ where $\mathbf{Q}^C$ is the set of irrational numbers, then X is *not* Riemann-integrable but we still have $\mathbf{E}(X) \equiv \int X d\mathbf{P} = 1$ being perfectly well-defined.

We sometimes call the expected value of X, with respect to the Lebesgue measure probability triple, its *Lebesgue integral*, and if $\mathbf{E}|X| < \infty$ we say that X is *Lebesgue integrable*. By Theorem 4.4.1, the Lebesgue integral is a *generalisation* of the Riemann integral. Hence, in addition to learning many things about probability theory, we are also learning more about integration at the same time!

We also note that we do not need to restrict attention to integrals over the unit interval $[0, 1]$. Indeed, if $X : \mathbf{R} \to [0, \infty)$ is Borel-measurable, then we can define $\int_{-\infty}^{\infty} X \, d\lambda$, where λ stands for Lebesgue measure on the entire real line $\mathbf{R}$, by

$$\int_{-\infty}^{\infty} X(t) \, \lambda(dt) = \sum_{n \in \mathbf{Z}} \int_0^1 X(n + t) \, \mathbf{P}(dt), \qquad (4.4.2)$$

or equivalently

$$\int_{-\infty}^{\infty} X(t) \, \lambda(dt) = \sum_{n \in \mathbf{Z}} \mathbf{E}(Y_n),$$

where $Y_n(t) = X(n + t)$ and the expectation is with respect to Lebesgue measure on $[0, 1]$. That is, we can integrate a non-negative function over the entire real line by adding up its integral over each interval $[n, n + 1)$. (For general X, we can then write $X = X^+ - X^-$, as usual.) In other words, we can represent λ, Lebesgue measure on $\mathbf{R}$, as a *sum* of countably many copies of Lebesgue measure on unit intervals. We shall see in Section 6 that we may use Lebesgue measure on $\mathbf{R}$ to help us define the distributions of general absolutely-continuous random variables.

Remark 4.4.3.   Measures like λ above, which are not finite but which can be written as the countable sum of finite measures, are called *σ-finite measures*. They are not probability measures, and cannot be used to define probability spaces; however by countable additivity they still satisfy many properties that probability measures do. Of course, unlike in the probability measure case, $\int_{-\infty}^{\infty} X \, d\lambda$ may be infinite or undefined even if X is bounded.

4.5. Exercises.

Exercise 4.5.1.   Let $(\Omega, \mathcal{F}, P)$ be Lebesgue measure on $[0, 1]$, and set

$$X(\omega) = \begin{cases} 1 \,, & 0 \le \omega < 1/4 \\ 2\omega^2, & 1/4 \le \omega < 3/4 \\ \omega^2, & 3/4 \le \omega \le 1. \end{cases}$$

Compute $P(X \in A)$ where
(a) $A = [0, 1]$.
(b) $A = [\frac{1}{2}, 1]$.

Exercise 4.5.2.   Let X be a random variable with finite mean, and let $a \in \mathbf{R}$ be any real number. Prove that $\mathbf{E}\big(\max(X, a)\big) \ge \max\big(\mathbf{E}(X), a\big)$. [Hint: Consider separately the cases $\mathbf{E}(X) \ge a$ and $\mathbf{E}(X) < a$.] (See also Exercise 5.5.7.)

Exercise 4.5.3.   Give an example of random variables X and Y defined on Lebesgue measure on $[0, 1]$, such that $P(X > Y) > \frac{1}{2}$, but $\mathbf{E}(X) < \mathbf{E}(Y)$.

Exercise 4.5.4.   Let $(\Omega, \mathcal{F}, \mathbf{P})$ be the uniform distribution on $\Omega = \{1, 2, 3\}$, as in Example 2.2.2. Find random variables X, Y, and Z on $(\Omega, \mathcal{F}, \mathbf{P})$ such that $\mathbf{P}(X > Y)\,\mathbf{P}(Y > Z)\,\mathbf{P}(Z > X) > 0$, and $\mathbf{E}(X) = \mathbf{E}(Y) = \mathbf{E}(Z)$.

Exercise 4.5.5.   Let X be a random variable on $(\Omega, \mathcal{F}, \mathbf{P})$, and suppose that Ω is a finite set. Prove that X is a simple random variable.

Exercise 4.5.6.   Let X be a random variable defined on Lebesgue measure on $[0, 1]$, and suppose that X is a one-to-one function, i.e. that if $\omega_1 \ne \omega_2$ then $X(\omega_1) \ne X(\omega_2)$. Prove that X is *not* a simple random variable.

Exercise 4.5.7. *(Principle of inclusion-exclusion, general case)* Let $A_1, A_2, \ldots, A_n \in \mathcal{F}$. Generalise the principle of inclusion-exclusion to:

$$\mathbf{P}(A_1 \cup \ldots \cup A_n) = \sum_{i=1}^{n} \mathbf{P}(A_i) - \sum_{1 \leq i < j \leq n} \mathbf{P}(A_i \cap A_j)$$

$$+ \sum_{1 \leq i < j < k \leq n} \mathbf{P}(A_i \cap A_j \cap A_k) - \ldots \pm \mathbf{P}(A_1 \cap \ldots \cap A_n).$$

[Hint: Expand $1 - \prod_{i=1}^{n}(1 - \mathbf{1}_{A_i})$, and take expectations of both sides.]

Exercise 4.5.8. Let $f(x) = ax^2 + bx + c$ be a second-degree polynomial function (where $a, b, c \in \mathbf{R}$ are constants).
(a) Find necessary and sufficient conditions on a, b, and c such that the equation $\mathbf{E}(f(\alpha X)) = \alpha^2 \mathbf{E}(f(X))$ holds for all $\alpha \in \mathbf{R}$ and all random variables X.
(b) Find necessary and sufficient conditions on a, b, and c such that the equation $\mathbf{E}(f(X - \beta)) = \mathbf{E}(f(X))$ holds for all $\beta \in \mathbf{R}$ and all random variables X.
(c) Do parts (a) and (b) account for the properties of the variance function? Why or why not?

Exercise 4.5.9. In proving property (4.1.6) of variance, why did we not simply proceed by induction on n? That is, suppose we know that $\mathbf{Var}(X + Y) = \mathbf{Var}(X) + \mathbf{Var}(Y)$ whenever X and Y are independent. Does it follow easily that $\mathbf{Var}(X + Y + Z) = \mathbf{Var}(X) + \mathbf{Var}(Y) + \mathbf{Var}(Z)$ whenever X, Y, and Z are independent? Why or why not? How does Exercise 3.6.6 fit in?

Exercise 4.5.10. Let $X_1, X_2, \ldots$ be i.i.d. with mean μ and variance σ^2, and let N be an integer-valued random variable with mean m and variance v, with N independent of all the X_i. Let $S = X_1 + \ldots + X_N = \sum_{i=1}^{\infty} X_i \mathbf{1}_{N \geq i}$. Compute $\mathbf{Var}(S)$ in terms of μ, σ^2, m, and v.

Exercise 4.5.11. Let X and Z be independent, each with the standard normal distribution, let $a, b \in \mathbf{R}$ (not both 0), and let $Y = aX + bZ$.
(a) Compute $\mathbf{Corr}(X, Y)$.
(b) Show that $|\mathbf{Corr}(X, Y)| \leq 1$ in this case. (Compare Exercise 5.5.6.)
(c) Give necessary and sufficient conditions on the values of a and b such that $\mathbf{Corr}(X, Y) = 1$.
(d) Give necessary and sufficient conditions on the values of a and b such that $\mathbf{Corr}(X, Y) = -1$.

Exercise 4.5.12.   Let X and Y be independent general non-negative random variables, and let $X_n = \Psi_n(X)$, where $\Psi_n(x) = \min(n, 2^{-n}\lfloor 2^n x \rfloor)$ as in Proposition 4.2.5.
(a) Give an example of a sequence of functions $\Phi_n : [0, \infty) \to [0, \infty)$, *other* than $\Phi_n(x) = \Psi_n(x)$, such that for all x, $0 \le \Phi_n(x) \le x$ and $\{\Phi_n(x)\} \nearrow x$ as $n \to \infty$.
(b) Suppose $Y_n = \Phi_n(Y)$ with Φ_n as in part (a). Must X_n and Y_n be independent?
(c) Suppose $\{Y_n\}$ is an arbitrary collection of non-negative simple random variables such that $\{Y_n\} \nearrow Y$. Must X_n and Y_n be independent?
(d) Under the assumption of part (c), determine (with proof) which quantities in equation (4.2.7) are necessarily equal.

Exercise 4.5.13.   Give examples of a random variable X defined on Lebesgue measure on $[0, 1]$, such that
(a) $\mathbf{E}(X^+) = \infty$ and $0 < \mathbf{E}(X^-) < \infty$.
(b) $\mathbf{E}(X^-) = \infty$ and $0 < \mathbf{E}(X^+) < \infty$.
(c) $\mathbf{E}(X^+) = \mathbf{E}(X^-) = \infty$.
(d) $\mathbf{E}(X) < \infty$ but $\mathbf{E}(X^2) = \infty$.

Exercise 4.5.14.   Let $Z_1, Z_2, \ldots$ be general random variables with $\mathbf{E}|Z_i| < \infty$, and let $Z = Z_1 + Z_2 + \ldots$.
(a) Suppose $\sum_i \mathbf{E}(Z_i^+) < \infty$ and $\sum_i \mathbf{E}(Z_i^-) < \infty$. Prove that $\mathbf{E}(Z) = \sum_i \mathbf{E}(Z_i)$.
(b) Show that we still have $\mathbf{E}(Z) = \sum_i \mathbf{E}(Z_i)$ if we have at least one of $\sum_i \mathbf{E}(Z_i^+) < \infty$ or $\sum_i \mathbf{E}(Z_i^-) < \infty$.
(c) Let $\{Z_i\}$ be independent, with $\mathbf{P}(Z_i = +1) = \mathbf{P}(Z_i = -1) = \frac{1}{2}$ for each i. Does $\mathbf{E}(Z) = \sum_i \mathbf{E}(Z_i)$ in this case? How does that relate to (4.2.8)?

Exercise 4.5.15.   Let $(\Omega_1, \mathcal{F}_1, \mathbf{P}_1)$ and $(\Omega_2, \mathcal{F}_2, \mathbf{P}_2)$ be two probability triples. Let $A_1, A_2, \ldots \in \mathcal{F}_1$, and $B_1, B_2, \ldots \in \mathcal{F}_2$. Suppose that it happens that the sets $\{A_n \times B_n\}$ are all disjoint, and furthermore that $\bigcup_{n=1}^{\infty}(A_n \times B_n) = A \times B$ for some $A \in \mathcal{F}_1$ and $B \in \mathcal{F}_2$.
(a) Prove that for each $\omega \in \Omega_1$, we have

$$1_A(\omega)\,\mathbf{P}_2(B) = \sum_{n=1}^{\infty} 1_{A_n}(\omega)\,\mathbf{P}_2(B_n).$$

[Hint: This is essentially countable additivity of $\mathbf{P}_2$, but you do need to be careful about disjointness.]
(b) By taking expectations of both sides with respect to $\mathbf{P}_1$ and using countable additivity of $\mathbf{P}_1$, prove that

$$\mathbf{P}_1(A)\,\mathbf{P}_2(B) = \sum_{n=1}^{\infty} \mathbf{P}_1(A_n)\,\mathbf{P}_2(B_n).$$

(c) Use this result to prove that the $\mathcal{J}$ and $\mathbf{P}$ for product measure, presented in Subsection 2.6, do indeed satisfy (2.5.5).

4.6. Section summary.

In this section we defined the expected value $\mathbf{E}(X)$ of a random variable X, first for simple random variables and then for general random variables. We proved basic properties such as linearity and order-preserving. We defined the variance $\mathbf{Var}(X)$. If X and Y are independent, then $\mathbf{E}(XY) = \mathbf{E}(X)\mathbf{E}(Y)$ and $\mathbf{Var}(X + Y) = \mathbf{Var}(X) + \mathbf{Var}(Y)$. We also proved the monotone convergence theorem. Finally, we connected expected value to Riemann (calculus-style) integration.

5. Inequalities and convergence.

In this section we consider various relationships regarding expected values and limits.

5.1. Various inequalities.

We begin with two very important inequalities about random variables.

Proposition 5.1.1. *(Markov's inequality.) If X is a non-negative random variable, then for all $\alpha > 0$,*

$$\mathbf{P}(X \geq \alpha) \leq \mathbf{E}(X)/\alpha.$$

In words, the probability that X exceeds α is bounded above by its mean divided by α.

Proof. Define a new random variable Z by

$$Z(\omega) = \begin{cases} \alpha, & X(\omega) \geq \alpha \\ 0, & X(\omega) < \alpha. \end{cases}$$

Then clearly $Z \leq X$, so that $\mathbf{E}(Z) \leq \mathbf{E}(X)$ by the order-preserving property. On the other hand, we compute that $\mathbf{E}(Z) = \alpha \, \mathbf{P}(X \geq \alpha)$. Hence, $\alpha \, \mathbf{P}(X \geq \alpha) \leq \mathbf{E}(X)$. ∎

Markov's inequality applies only to non-negative random variables, but it immediately implies another inequality which holds more generally:

Proposition 5.1.2. *(Chebychev's inequality.) Let Y be an arbitrary random variable, with finite mean μ_Y. Then for all $\alpha > 0$,*

$$\mathbf{P}\left(|Y - \mu_Y| \geq \alpha\right) \leq \mathbf{Var}(Y)/\alpha^2.$$

In words, the probability that Y differs from its mean by more than α is bounded above by its variance divided by α^2.

Proof. Set $X = (Y - \mu_Y)^2$. Then X is a non-negative random variable. Thus, using Proposition 5.1.1, we have

$$\mathbf{P}\left(|Y - \mu_y| \geq \alpha\right) = \mathbf{P}\left(X \geq \alpha^2\right) \leq \mathbf{E}(X)/\alpha^2 = \mathbf{Var}(Y)/\alpha^2. \qquad ∎$$

We shall use the above two inequalities extensively, including to prove the laws of large numbers presented below. Two other sometimes-useful inequalities are as follows.

Proposition 5.1.3.   *(Cauchy-Schwarz inequality.) Let X and Y be random variables with $\mathbf{E}(X^2) < \infty$ and $\mathbf{E}(Y^2) < \infty$. Then $\mathbf{E}|XY| \leq \sqrt{\mathbf{E}(X^2)\,\mathbf{E}(Y^2)}$.*

Proof.   Let $Z = |X| \,/\, \sqrt{\mathbf{E}(X^2)}$ and $W = |Y| \,/\, \sqrt{\mathbf{E}(Y^2)}$, so that $\mathbf{E}(Z^2) = \mathbf{E}(W^2) = 1$. Then

$$0 \;\leq\; \mathbf{E}((Z - W)^2) \;=\; \mathbf{E}(Z^2 + W^2 - 2ZW) \;=\; 1 + 1 - 2\,\mathbf{E}(ZW),$$

so $\mathbf{E}(ZW) \leq 1$, i.e. $\mathbf{E}|XY| \leq \sqrt{\mathbf{E}(X^2)\,\mathbf{E}(Y^2)}$.   ∎

Proposition 5.1.4.   *(Jensen's inequality.) Let X be a random variable with finite mean, and let $\phi : \mathbf{R} \to \mathbf{R}$ be a convex function, i.e. a function such that $\lambda\phi(x) + (1 - \lambda)\phi(y) \geq \phi(\lambda x + (1 - \lambda)y)$ for $x, y, \lambda \in \mathbf{R}$ and $0 \leq \lambda \leq 1$. Then $\mathbf{E}\,(\phi(X)) \geq \phi\,(\mathbf{E}(X))$.*

Proof.   Since ϕ is convex, we can find a linear function $g(x) = ax + b$ which lies entirely below the graph of ϕ but which touches it at the point $x = \mathbf{E}(X)$, i.e. such that $g(x) \leq \phi(x)$ for all $x \in \mathbf{R}$, and $g(\mathbf{E}(X)) = \phi(\mathbf{E}(X))$. Then

$$\mathbf{E}(\phi(X)) \geq \mathbf{E}(g(X)) = \mathbf{E}(aX + b) = a\mathbf{E}(X) + b = g(\mathbf{E}(X)) = \phi(\mathbf{E}(X)).\ \blacksquare$$

5.2. Convergence of random variables.

If $Z, Z_1, Z_2, \ldots$ are random variables defined on some $(\Omega, \mathcal{F}, \mathbf{P})$, what does it mean to say that $\{Z_n\}$ *converges* to Z as $n \to \infty$?

One notion we have already seen (cf. Theorem 4.2.2) is *pointwise convergence*, i.e. $\lim_{n\to\infty} Z_n(\omega) = Z(\omega)$. A slightly weaker notion which often arises is convergence *almost surely* (or, *a.s.* or *with probability 1* or *w.p. 1* or *almost everywhere*), meaning that $\mathbf{P}(\lim_{n\to\infty} Z_n = Z) = 1$, i.e. that $\mathbf{P}\{\omega \in \Omega : \lim_{n\to\infty} Z_n(\omega) = Z(\omega)\} = 1$. As an aid to establishing such convergence, we have the following:

Lemma 5.2.1.   *Let $Z, Z_1, Z_2, \ldots$ be random variables. Suppose for each $\epsilon > 0$, we have $\mathbf{P}(|Z_n - Z| \geq \epsilon \ i.o.) = 0$. Then $\mathbf{P}(Z_n \to Z) = 1$, i.e. $\{Z_n\}$ converges to Z almost surely.*

Proof.   It follows from Proposition A.3.3 that

$$\mathbf{P}(Z_n \to Z) = \mathbf{P}(\forall\, \epsilon > 0,\ |Z_n - Z| < \epsilon\ a.a.) = 1 - \mathbf{P}(\exists\, \epsilon > 0,\ |Z_n - Z| \geq \epsilon\ i.o.)\,.$$

By countable subadditivity, we have that

$$\mathbf{P}(\exists\, \epsilon > 0,\ \epsilon\ \text{rational},\ |Z_n - Z| \geq \epsilon\ i.o.) \leq \sum_{\substack{\epsilon > 0 \\ \epsilon\ \text{rational}}} \mathbf{P}(|Z_n - Z| \geq \epsilon\ i.o.) = 0\,.$$

But given *any* $\epsilon > 0$, there exists a rational $\epsilon' > 0$ with $\epsilon' < \epsilon$. For this ϵ', we have that $\{|Z_n - Z| \geq \epsilon\ i.o.\} \subseteq \{|Z_n - Z| \geq \epsilon'\ i.o.\}$. It follows that

$$\mathbf{P}(\exists\, \epsilon > 0,\ |Z_n - Z| \geq \epsilon\, i.o.) \leq \mathbf{P}(\exists\, \epsilon' > 0,\ \epsilon'\ \text{rational},\ |Z_n - Z| \geq \epsilon'\, i.o.) = 0\,,$$

thus giving the result. ∎

Combining Lemma 5.2.1 with the Borel-Cantelli Lemma, we obtain:

Corollary 5.2.2. *Let $Z, Z_1, Z_2, \ldots$ be random variables. Suppose for each $\epsilon > 0$, we have $\sum_n \mathbf{P}(|Z_n - Z| \geq \epsilon) < \infty$. Then $\mathbf{P}(Z_n \to Z) = 1$, i.e. $\{Z_n\}$ converges to Z almost surely.*

Another notion only involves probabilities: we say that $\{Z_n\}$ converges to Z *in probability* if for all $\epsilon > 0$, $\lim_{n\to\infty} \mathbf{P}(|Z_n - Z| \geq \epsilon) = 0$. We next consider the relation between convergence in probability, and convergence almost surely.

Proposition 5.2.3. *Let $Z, Z_1, Z_2, \ldots$ be random variables. Suppose $Z_n \to Z$ almost surely (i.e., $\mathbf{P}(Z_n \to Z) = 1$). Then $Z_n \to Z$ in probability (i.e., for any $\epsilon > 0$, we have $\mathbf{P}(|Z_n - Z| \geq \epsilon) \to 0$). That is, if a sequence of random variables converges almost surely, then it converges in probability to the same limit.*

Proof. Fix $\epsilon > 0$, and let $A_n = \{\omega;\ \exists\, m \geq n,\ |Z_m - Z| \geq \epsilon\}$. Then $\{A_n\}$ is a decreasing sequence of events. Furthermore, if $\omega \in \bigcap_n A_n$, then $Z_n(\omega) \not\to Z(\omega)$ as $n \to \infty$. Hence, $\mathbf{P}(\bigcap_n A_n) \leq \mathbf{P}(Z_n \not\to Z) = 0$. By continuity of probabilities, we have $\mathbf{P}(A_n) \to \mathbf{P}(\bigcap_n A_n) = 0$. Hence, $\mathbf{P}(|Z_n - Z| \geq \epsilon) \leq \mathbf{P}(A_n) \to 0$, as required. ∎

On the other hand, the converse to Proposition 5.2.3 is false. (This justifies the use of "strong" and "weak" in describing the two Laws of Large Numbers below.) For a first example, let $\{Z_n\}$ be independent, with $\mathbf{P}(Z_n = 1) = 1/n = 1 - \mathbf{P}(Z_n = 0)$. (Formally, the existence of such $\{Z_n\}$ follows from Theorem 7.1.1.) Then clearly Z_n converges to 0 in probability. On the other hand, by the Borel-Cantelli Lemma, $\mathbf{P}(Z_n = 1\ i.o.) = 1$, so $\mathbf{P}(Z_n \to 0) = 0$, so Z_n does *not* converge to 0 almost surely.

For a second example, let $(\Omega, \mathcal{F}, \mathbf{P})$ be Lebesgue measure on $[0, 1]$, and set $Z_1 = \mathbf{1}_{[0, \frac{1}{2})}$, $Z_2 = \mathbf{1}_{[\frac{1}{2}, 1]}$, $Z_3 = \mathbf{1}_{[0, \frac{1}{4})}$, $Z_4 = \mathbf{1}_{[\frac{1}{4}, \frac{1}{2})}$, $Z_5 = \mathbf{1}_{[\frac{1}{2}, \frac{3}{4})}$, $Z_6 =$

$1_{[\frac{3}{4},1]}$, $Z_7 = 1_{[0,\frac{1}{8})}$, $Z_8 = 1_{[\frac{1}{8},\frac{1}{4})}$, etc. Then, by inspection, Z_n converges to 0 in probability, but Z_n does *not* converge to 0 almost surely.

5.3. Laws of large numbers.

Here we prove a first form of the weak law of large numbers.

Theorem 5.3.1.   (*Weak law of large numbers – first version.*) *Let* $X_1, X_2, \ldots$ *be a sequence of independent random variables, each having the same mean* m, *and each having variance* $\leq v < \infty$. *Then for all* $\epsilon > 0$,

$$\lim_{n \to \infty} \mathbf{P}\left(\left|\frac{1}{n}(X_1 + X_2 + \ldots + X_n) - m\right| \geq \epsilon\right) = 0.$$

In words, the partial averages $\frac{1}{n}(X_1 + X_2 + \ldots + X_n)$ *converge in probability to* m.

Proof.   Set $S_n = \frac{1}{n}(X_1 + X_2 + \ldots + X_n)$. Then using linearity of expected value, and also properties (4.1.5) and (4.1.6) of variance, we see that $\mathbf{E}(S_n) = m$ and $\mathbf{Var}(S_n) \leq v/n$. Hence by Chebychev's inequality (Theorem 5.1.2), we have

$$\mathbf{P}\left(\left|\frac{1}{n}(X_1 + X_2 + \ldots + X_n) - m\right| \geq \epsilon\right) \leq v/\epsilon^2 n \to 0, \qquad n \to \infty,$$

as required.   ∎

For example, in the case of infinite coin tossing, if $X_i = r_i = 0$ or 1 as the ith coin is tails or heads, then Theorem 5.3.1 states that the probability that the fraction of heads on the first n tosses differs from $\frac{1}{2}$ by more than ϵ goes to 0 as $n \to \infty$. Informally, the fraction of heads gets closer and closer to $\frac{1}{2}$ with higher and higher probability.

We next prove a first form of the strong law of large numbers.

Theorem 5.3.2.   (*Strong law of large numbers – first version.*) *Let* $X_1, X_2, \ldots$ *be a sequence of independent random variables, each having the same finite mean* m, *and each having* $\mathbf{E}((X_i - m)^4) \leq a < \infty$. *Then*

$$\mathbf{P}\left(\lim_{n \to \infty} \frac{1}{n}(X_1 + X_2 + \ldots + X_n) = m\right) = 1.$$

In words, the partial averages $\frac{1}{n}(X_1 + X_2 + \ldots + X_n)$ *converge almost surely to* m.

Proof. Since $(X_i - m)^2 \leq (X_i - m)^4 + 1$ (consider separately the two cases $(X_i - m)^2 \leq 1$ and $(X_i - m)^2 > 1$), it follows that each X_i has variance $\leq a + 1 \equiv v < \infty$. We assume for simplicity that $m = 0$; if not then we can simply replace X_i by $X_i - m$.

Let $S_n = X_1 + X_2 + \ldots + X_n$, and consider $\mathbf{E}(S_n^4)$. Now, S_n^4 will (when multiplied out) contain many terms of the form $X_i X_j X_k X_\ell$ for i, j, k, ℓ distinct, but all of these have expected value zero. Similarly, it will contain many terms of the form $X_i X_j (X_k)^2$ and $X_i (X_j)^3$ which also have expected value zero. The only terms with non-vanishing expectation will be n terms of the form $(X_i)^4$, and $\binom{n}{2}\binom{4}{2} = 3n(n-1)$ terms of the form $(X_i)^2 (X_j)^2$ with $i \neq j$. Now, $\mathbf{E}\left((X_i)^4\right) \leq a$. Furthermore, if $i \neq j$ then X_i^2 and X_j^2 are independent by Proposition 3.2.3, so since $m = 0$ we have $\mathbf{E}\left((X_i)^2 (X_j)^2\right) = \mathbf{E}\left((X_i)^2\right)\mathbf{E}\left((X_j)^2\right) = \mathbf{Var}(X_i)\mathbf{Var}(X_j) \leq v^2$. We conclude that $\mathbf{E}(S_n^4) \leq na + 3n(n-1)v^2 \leq Kn^2$ where $K = a + 3v^2$. This is the key.

To finish, we note that for any $\epsilon > 0$, we have by Markov's inequality that

$$\mathbf{P}\left(\left|\frac{1}{n} S_n\right| \geq \epsilon\right) = \mathbf{P}\left(|S_n| \geq n\epsilon\right) = \mathbf{P}\left(|S_n|^4 \geq n^4 \epsilon^4\right)$$

$$\leq \mathbf{E}\left(S_n^4\right)/(n^4 \epsilon^4) \leq Kn^2/(n^4 \epsilon^4) = K\epsilon^{-4}\frac{1}{n^2}.$$

Since $\sum_{n=1}^{\infty} \frac{1}{n^2} < \infty$ by (A.3.7), it follows from Corollary 5.2.2 that $\frac{1}{n} S_n$ converges to 0 almost surely. ■

For example, in the case of coin tossing, Theorem 5.3.2 states that the fraction of heads on the first n tosses will *converge*, as $n \to \infty$, to $\frac{1}{2}$. Although this conclusion sounds quite similar to the corresponding conclusion from Theorem 5.3.1, we know from Proposition 5.2.3 (and the examples following) that it is actually a stronger result.

5.4. Eliminating the moment conditions.

Theorems 5.3.1 and 5.3.2 provide clear evidence that the partial sums $\frac{1}{n}(X_1 + \ldots + X_n)$ are indeed converging to the common mean m in some sense. However, they require that the variance (i.e., second moment) or even the fourth moment of the X_i be finite (and uniformly bounded). This is an unnatural condition which is sometimes difficult to check.

Thus, in this section we develop a new form of the strong law of large numbers which requires only that the mean (i.e., first moment) of the random variables be finite. However, as a penalty, it demands that the random variables be "i.i.d." as opposed to merely independent. (Of course, once we have proven a strong law, we will immediately obtain a weak law by using

Proposition 5.2.3.) Our proof follows the approach in Billingsley (1995). We begin with a definition.

Definition 5.4.1.   A collection of random variables $\{X_\alpha\}_{\alpha \in I}$ are *identically distributed* if for any Borel-measurable function $f : \mathbf{R} \to \mathbf{R}$, the expected value $\mathbf{E}(f(X_\alpha))$ does not depend on α, i.e. is the same for all $\alpha \in I$.

Remark 5.4.2.   It follows from Proposition 6.0.2 and Corollary 6.1.3 below that $\{X_\alpha\}_{\alpha \in I}$ are identically distributed if and only if for all $x \in \mathbf{R}$, the probability $\mathbf{P}(X_\alpha \le x)$ does not depend on α.

Definition 5.4.3.   A collection of random variables $\{X_\alpha\}_{\alpha \in I}$ are *i.i.d.* if they are independent and are also identically distributed.

Theorem 5.4.4.   *(Strong law of large numbers – second version.) Let* $X_1, X_2, \ldots$ *be a sequence of i.i.d. random variables, each having finite mean* m. *Then*

$$\mathbf{P}\left(\lim_{n \to \infty} \frac{1}{n}(X_1 + X_2 + \ldots + X_n) = m\right) = 1.$$

In words, the partial averages $\frac{1}{n}(X_1 + X_2 + \ldots + X_n)$ *converge almost surely to* m.

Proof.   The proof is somewhat difficult because we do not assume the finiteness of any moments of the X_i other than the first. Instead, we shall use a "truncation argument", by defining new random variables Y_i which are truncated versions of the corresponding X_i. Thus, higher moments will exist for the Y_i, even though the Y_i will tend to be "similar" to the X_i.

To begin, we assume that $X_i \ge 0$; if not, we can consider separately X_i^+ and X_i^-. (Note that we *cannot* now assume without loss of generality that $m = 0$.) We set $Y_i = X_i \mathbf{1}_{X_i \le i}$, i.e. $Y_i = X_i$ unless X_i exceeds i, in which case $Y_i = 0$. Then since $0 \le Y_i \le i$, therefore $\mathbf{E}(Y_i^k) \le i^k < \infty$. Also, by Proposition 3.2.3, the $\{Y_i\}$ are independent. Furthermore, since the X_i are i.i.d., we have that $\mathbf{E}(Y_i) = \mathbf{E}(X_i \mathbf{1}_{X_i \le i}) = \mathbf{E}(X_1 \mathbf{1}_{X_1 \le i}) \nearrow \mathbf{E}(X_1) = m$ as $i \to \infty$, by the monotone convergence theorem.

We set $S_n = \sum_{i=1}^{n} X_i$, and set $S_n^* = \sum_{i=1}^{n} Y_i$. We compute using (4.1.6) and (4.1.4) that

$$\mathbf{Var}(S_n^*) = \mathbf{Var}(Y_1) + \ldots + \mathbf{Var}(Y_n) \le \mathbf{E}(Y_1^2) + \ldots + \mathbf{E}(Y_n^2)$$

$$= \mathbf{E}(X_1^2 \mathbf{1}_{X_1 \le 1}) + \ldots + \mathbf{E}(X_n^2 \mathbf{1}_{X_n \le n}) \le n\mathbf{E}(X_1^2 \mathbf{1}_{X_1 \le n}) \le n^3 < \infty. \quad (5.4.5)$$

We now choose $\alpha > 1$ (we will later let $\alpha \searrow 1$), and set $u_n = \lfloor \alpha^n \rfloor$, i.e. u_n is the greatest integer not exceeding α^n. Then $u_n \le \alpha^n$. Furthermore, since $\alpha^n > 1$, it follows (consider separately the cases $\alpha^n < 2$ and $\alpha^n \geq 2$) that $u_n \geq \alpha^n/2$, i.e. $1/u_n \leq 2/\alpha^n$. Hence, for any $x > 0$, we have that

$$\sum_{\substack{n \\ u_n \geq x}} 1/u_n \leq \sum_{\substack{n \\ \alpha^n \geq x}} 1/u_n \leq \sum_{\substack{n \\ \alpha^n \geq x}} 2/\alpha^n \leq \sum_{k=\log_\alpha x}^{\infty} 2/\alpha^k = \frac{2/x}{1 - \frac{1}{\alpha}}. \qquad (5.4.6)$$

(Note that here we sum over $k = \log_\alpha x, \log_\alpha x + 1, \ldots$, even if $\log_\alpha x$ is not an integer.)

We now proceed to the heart of the proof. For any $\epsilon > 0$, we have that

$$
\begin{aligned}
& \sum_{n=1}^{\infty} \mathbf{P}\left(\left| \frac{S_{u_n}^* - \mathbf{E}(S_{u_n}^*)}{u_n} \right| \geq \epsilon \right) \\
= \; & \sum_{n=1}^{\infty} \mathbf{P}\left(\left| \frac{S_{u_n}^*}{u_n} - \mathbf{E}(\frac{S_{u_n}^*}{u_n}) \right| \geq \epsilon \right) \\
\leq \; & \sum_{n=1}^{\infty} \frac{\mathrm{Var}(S_{u_n}^*/u_n)}{\epsilon^2} && \text{by Chebychev's inequality} \\
= \; & \sum_{n=1}^{\infty} \frac{\mathrm{Var}(S_{u_n}^*)}{u_n^2 \epsilon^2} && \text{by (4.1.5)} \\
\leq \; & \sum_{n=1}^{\infty} \frac{u_n \mathbf{E}(X_1^2 \mathbf{1}_{X_1 \leq u_n})}{u_n^2 \epsilon^2} && \text{by (5.4.5)} \\
= \; & \frac{1}{\epsilon^2} \mathbf{E}\left(X_1^2 \sum_{n=1}^{\infty} \frac{1}{u_n} \mathbf{1}_{u_n \geq X_1} \right) && \text{by countable linearity} \\
\leq \; & \frac{1}{\epsilon^2} \mathbf{E}\left(X_1^2 \frac{2/X_1}{1 - \frac{1}{\alpha}} \right) && \text{by (5.4.6)} \\
= \; & \frac{2}{\epsilon^2 (1 - \frac{1}{\alpha})} \mathbf{E}(X_1) \\
= \; & \frac{2m}{\epsilon^2 (1 - \frac{1}{\alpha})} < \infty.
\end{aligned}
$$

This finiteness is the key. It now follows from Corollary 5.2.2 that

$$\left\{ \frac{S_{u_n}^* - \mathbf{E}(S_{u_n}^*)}{u_n} \right\} \text{ converges to 0 almost surely}. \qquad (5.4.7)$$

To complete the proof, we need to replace $\frac{\mathbf{E}(S_{u_n}^*)}{u_n}$ by m, and replace $S_{u_n}^*$ by S_{u_n}, and finally replace the index u_n by the general index k. We consider each of these three issues in turn.

First, since $\mathbf{E}(Y_i) \to m$ as $i \to \infty$, and since $u_n \to \infty$ as $n \to \infty$, it follows immediately that $\frac{\mathbf{E}(S_{u_n}^*)}{u_n} \to m$ as $n \to \infty$. Hence, it follows from (5.4.7) that

$$\left\{ \frac{S_{u_n}^*}{u_n} \right\} \text{ converges to } m \text{ almost surely}. \qquad (5.4.8)$$

Second, we note by Proposition 4.2.9 that

$$\sum_{k=1}^{\infty} \mathbf{P}(X_k \neq Y_k) = \sum_{k=1}^{\infty} \mathbf{P}(X_k > k) \leq \sum_{k=1}^{\infty} \mathbf{P}(X_k \geq k)$$

$$= \sum_{k=1}^{\infty} \mathbf{P}(X_1 \geq k) = \mathbf{E}\lfloor X_1 \rfloor \leq \mathbf{E}(X_1) = m < \infty.$$

Hence, again by Borel-Cantelli, we have that $\mathbf{P}(X_k \neq Y_k \ i.o.) = 0$, so that $\mathbf{P}(X_k = Y_k \ a.a.) = 1$. It follows that, with probability one, as $n \to \infty$ the limit of $\frac{S^*_{u_n}}{u_n}$ coincides with the limit of $\frac{S_{u_n}}{u_n}$. Hence, (5.4.8) implies that

$$\left\{ \frac{S_{u_n}}{u_n} \right\} \text{ converges to } m \text{ almost surely}. \tag{5.4.9}$$

Finally, for an arbitrary index k, we can find $n = n_k$ so that $u_n \leq k < u_{n+1}$. But then

$$\frac{u_n}{u_{n+1}} \frac{S_{u_n}}{u_n} = \frac{S_{u_n}}{u_{n+1}} \leq \frac{S_k}{k} \leq \frac{S_{u_{n+1}}}{u_n} = \frac{S_{u_{n+1}}}{u_{n+1}} \frac{u_{n+1}}{u_n}. \tag{5.4.10}$$

Now, as $k \to \infty$, we have $n = n_k \to \infty$, so that $\frac{u_n}{u_{n+1}} \to \frac{1}{\alpha}$ and $\frac{u_{n+1}}{u_n} \to \alpha$. Hence, for any $\alpha > 1$ and $\delta > 0$, with probability 1 we have $m/(1+\delta)\alpha \leq \frac{S_k}{k} \leq (1+\delta)\alpha m$ for all sufficiently large k. For any $\epsilon > 0$, choosing $\alpha > 1$ and $\delta > 0$ so that $m/(1+\delta)\alpha > m - \epsilon$ and $(1+\delta)\alpha m < m + \epsilon$, this implies that $\mathbf{P}(|(S_k/k) - m| \geq \epsilon \ i.o.) = 0$. Hence, by Lemma 5.2.1, we have that as $k \to \infty$,

$$\frac{S_k}{k} \text{ converges to } m \text{ almost surely},$$

as required.   ∎

Using Proposition 5.2.3, we immediately obtain a corresponding statement about convergence in probability.

Corollary 5.4.11.   (*Weak law of large numbers – second version.*) *Let* $X_1, X_2, \ldots$ *be a sequence of i.i.d. random variables, each having finite mean* m. *Then for all* $\epsilon > 0$,

$$\lim_{n \to \infty} \mathbf{P}\left(\left| \frac{1}{n}(X_1 + X_2 + \ldots + X_n) - m \right| \geq \epsilon \right) = 0.$$

In words, the partial averages $\frac{1}{n}(X_1 + X_2 + \ldots + X_n)$ *converge in probability to* m.

5.5. Exercises.

Exercise 5.5.1. Suppose $\mathbf{E}(2^X) = 4$. Prove that $\mathbf{P}(X \geq 3) \leq 1/2$.

Exercise 5.5.2. Give an example of a random variable X and $\alpha > 0$ such that $\mathbf{P}(X \geq \alpha) > \mathbf{E}(X)/\alpha$. [Hint: Obviously X cannot be non-negative.] Where does the proof of Markov's inequality break down in this case?

Exercise 5.5.3. Give examples of random variables Y with mean 0 and variance 1 such that
(a) $\mathbf{P}(|Y| \geq 2) = 1/4$.
(b) $\mathbf{P}(|Y| \geq 2) < 1/4$.

Exercise 5.5.4. Suppose X is a non-negative random variable with $\mathbf{E}(X) = \infty$. What does Markov's inequality say in this case?

Exercise 5.5.5. Suppose Y is a random variable with finite mean μ_Y and with $\mathbf{Var}(Y) = \infty$. What does Chebychev's inequality say in this case?

Exercise 5.5.6. For general jointly defined random variables X and Y, prove that $|\mathbf{Corr}(X, Y)| \leq 1$. [Hint: Don't forget the Cauchy-Schwarz inequality.] (Compare Exercise 4.5.11.)

Exercise 5.5.7. Let $a \in \mathbf{R}$, and let $\phi(x) = \max(x, a)$ as in Exercise 4.5.2. Prove that ϕ is a convex function. Relate this to Jensen's inequality and to Exercise 4.5.2.

Exercise 5.5.8. Let $\phi(x) = x^2$.
(a) Prove that ϕ is a convex function.
(b) What does Jensen's inequality say for this choice of ϕ?
(c) Where in the text have we already seen the result of part (b)?

Exercise 5.5.9. Prove *Cantelli's inequality*, which states that if X is a random variable with finite mean m and finite variance v, then for $\alpha > 0$,

$$\mathbf{P}(X - m \geq \alpha) \leq \frac{v}{v + \alpha^2}.$$

[Hint: First show $\mathbf{P}(X - m \geq \alpha) \leq \mathbf{P}((X - m + y)^2 \geq (\alpha + y)^2)$. Then use Markov's inequality, and minimise the resulting bound over choice of $y > 0$.]

Exercise 5.5.10. Let $X_1, X_2, \ldots$ be a sequence of random variables, with $\mathbf{E}[X_n] = 8$ and $\mathbf{Var}[X_n] = 1/\sqrt{n}$ for each n. Prove or disprove that $\{X_n\}$ must converge to 8 in probability.

Exercise 5.5.11. Give (with proof) an example of a sequence $\{Y_n\}$ of jointly-defined random variables, such that as $n \to \infty$: (i) Y_n/n converges to 0 in probability; and (ii) Y_n/n^2 converges to 0 with probability 1; but (iii) Y_n/n does <u>not</u> converge to 0 with probability 1.

Exercise 5.5.12. Give (with proof) an example of two discrete random variables having the same mean and the same variance, but which are *not* identically distributed.

Exercise 5.5.13. Let $r \in \mathbf{N}$. Let $X_1, X_2, \ldots$ be identically distributed random variables having finite mean m, which are *r-dependent*, i.e. such that $X_{k_1}, X_{k_2}, \ldots, X_{k_j}$ are independent whenever $k_{i+1} > k_i + r$ for each i. (Thus, independent random variables are 0-dependent.) Prove that with probability one, $\frac{1}{n} \sum_{i=1}^{n} X_i \to m$ as $n \to \infty$. [Hint: Break up the sum $\sum_{i=1}^{n} X_i$ into r different sums.]

Exercise 5.5.14. Prove the converse of Lemma 5.2.1. That is, prove that if $\{X_n\}$ converges to X almost surely, then for each $\epsilon > 0$ we have $\mathbf{P}(|X_n - X| \geq \epsilon \ i.o.) = 0$.

Exercise 5.5.15. Let $X_1, X_2, \ldots$ be a sequence of independent random variables with $\mathbf{P}(X_n = 3^n) = \mathbf{P}(X_n = -3^n) = \frac{1}{2}$. Let $S_n = X_1 + \ldots + X_n$.
(a) Compute $\mathbf{E}(X_n)$ for each n.
(b) For $n \in \mathbf{N}$, compute $R_n \equiv \sup\{r \in \mathbf{R}; \ \mathbf{P}(|S_n| \geq r) = 1\}$, i.e. the largest number such that $|S_n|$ is always at least R_n.
(c) Compute $\lim_{n\to\infty} \frac{1}{n} R_n$.
(d) For which $\epsilon > 0$ (if any) is it the case that $\mathbf{P}(\frac{1}{n}|S_n| \geq \epsilon) \not\to 0$?
(e) Why does this result not contradict the various laws of large numbers?

5.6. Section summary.

This section presented inequalities about random variables. The first, Markov's inequality, provides an upper bound on $\mathbf{P}(X \geq \alpha)$ for non-negative random variables X. The second, Chebychev's inequality, provides an upper bound on $\mathbf{P}(|Y - \mu_Y| \geq \alpha)$ in terms of the variance of Y. The Cauchy-Schwarz and Jensen inequalities were also discussed.

It then discussed convergence of sequences of random variables, and presented various versions of the Law of Large Numbers. This law concerns partial averages $\frac{1}{n}(X_1 + \ldots + X_n)$ of collections of independent random variables $\{X_n\}$ all having the same mean m. Under the assumption of either finite higher moments (First Version) or identical distributions (Second Version), we proved that these partial averages converge in probability (Weak Law) or almost surely (Strong Law) to the mean m.

6. Distributions of random variables.

The *distribution* or *law* of a random variable is defined as follows.

Definition 6.0.1. *Given a random variable X on a probability triple $(\Omega, \mathcal{F}, \mathbf{P})$, its distribution (or law) is the function μ defined on $\mathcal{B}$, the Borel subsets of $\mathbf{R}$, by*

$$\mu(B) = \mathbf{P}(X \in B) = \mathbf{P}\left(X^{-1}(B)\right), \qquad B \in \mathcal{B}.$$

If μ is the law of a random variable, then $(\mathbf{R}, \mathcal{B}, \mu)$ is a valid probability triple. We shall sometimes write μ as $\mathcal{L}(X)$ or as $\mathbf{P}\,X^{-1}$. We shall also write $X \sim \mu$ to indicate that μ is the distribution of X.

We define the *cumulative distribution function* of a random variable X by $F_X(x) = \mathbf{P}(X \leq x)$, for $x \in \mathbf{R}$. By continuity of probabilities, the function F_X is *right-continuous*, i.e. if $\{x_n\} \searrow x$ then $F_X(x_n) \to F_X(x)$. It is also clearly a non-decreasing function of x, with $\lim_{x \to \infty} F_X(x) = 1$ and $\lim_{x \to -\infty} F_X(x) = 0$. We note the following.

Proposition 6.0.2. *Let X and Y be two random variables (possibly defined on different probability triples). Then $\mathcal{L}(X) = \mathcal{L}(Y)$ if and only if $F_X(x) = F_Y(x)$ for all $x \in \mathbf{R}$.*

Proof. The "if" part follows from Corollary 2.5.9. The "only if" part is immediate upon setting $B = (-\infty, x]$. ∎

6.1. Change of variable theorem.

The following result shows that distributions specify completely the expected values of random variables (and functions of them).

Theorem 6.1.1. *(Change of variable theorem.) Given a probability triple $(\Omega, \mathcal{F}, \mathbf{P})$, let X be a random variable having distribution μ. Then for any Borel-measurable function $f : \mathbf{R} \to \mathbf{R}$, we have*

$$\int_\Omega f\left(X(\omega)\right) \mathbf{P}(d\omega) = \int_{-\infty}^{\infty} f(t)\, \mu(dt), \qquad (6.1.2)$$

i.e. $\mathbf{E}_{\mathbf{P}}[f(X)] = \mathbf{E}_\mu(f)$, provided that either side is well-defined. In words, the expected value of the random variable $f(X)$ with respect to the probability measure $\mathbf{P}$ on Ω is equal to the expected value of the function f with respect to the measure μ on $\mathbf{R}$.

Proof.   Suppose first that $f = 1_B$ is an indicator function of a Borel set $B \subseteq \mathbf{R}$. Then $\int_\Omega f\left(X(\omega)\right)\mathbf{P}(d\omega) = \int_\Omega 1_{\{X(\omega)\in B\}}\mathbf{P}(d\omega) = \mathbf{P}(X \in B)$, while $\int_{-\infty}^{\infty} f(t)\mu(dt) = \int_{-\infty}^{\infty} 1_{\{t\in B\}}\mu(dt) = \mu(B) = \mathbf{P}(X \in B)$, so equality holds in this case.

Now suppose that f is a non-negative simple function. Then f is a finite positive linear combination of indicator functions. But since both sides of (6.1.2) are linear functions of f, we see that equality still holds in this case.

Next suppose that f is a general non-negative Borel-measurable function. Then by Proposition 4.2.5, we can find a sequence $\{f_n\}$ of non-negative simple functions such that $\{f_n\} \nearrow f$. We know that (6.1.2) holds when f is replaced by f_n. But then by letting $n \to \infty$ and using the Monotone Convergence Theorem (Theorem 4.2.2), we see that (6.1.2) holds for f as well.

Finally, for general Borel-measurable f, we can write $f = f^+ - f^-$. Since (6.1.2) holds for f^+ and for f^- separately, and since it is linear, therefore it must also hold for f.   ∎

Remark.   The method of proof used in Theorem 6.1.1 (namely considering first indicator functions, then non-negative simple functions, then general non-negative functions, and finally general functions) is quite widely applicable; we shall use it again in the next subsection.

Corollary 6.1.3.   *Let X and Y be two random variables (possibly defined on different probability triples). Then $\mathcal{L}(X) = \mathcal{L}(Y)$ if and only if $\mathbf{E}[f(X)] = \mathbf{E}[f(Y)]$ for all Borel-measurable $f : \mathbf{R} \to \mathbf{R}$ for which either expectation is well-defined. (Compare Proposition 6.0.2 and Remark 5.4.2.)*

Proof.   If $\mathcal{L}(X) = \mathcal{L}(Y) = \mu$ (say), then Theorem 6.1.1 says that $\mathbf{E}[f(X)] = \mathbf{E}[f(Y)] = \int_{\mathbf{R}} f\, d\mu$.

Conversely, if $\mathbf{E}[f(X)] = \mathbf{E}[f(Y)]$ for all Borel-measurable $f : \mathbf{R} \to \mathbf{R}$, then setting $f = 1_B$ shows that $\mathbf{P}[X \in B] = \mathbf{P}[Y \in B]$ for all Borel $B \subseteq \mathbf{R}$, i.e. that $\mathcal{L}(X) = \mathcal{L}(Y)$.   ∎

Corollary 6.1.4.   *If X and Y are random variables with $\mathbf{P}(X = Y) = 1$, then $\mathbf{E}[f(X)] = \mathbf{E}[f(Y)]$ for all Borel-measurable $f : \mathbf{R} \to \mathbf{R}$ for which either expectation is well-defined.*

Proof.   It follows directly that $\mathcal{L}(X) = \mathcal{L}(Y)$. Then, letting $\mu = \mathcal{L}(X) = \mathcal{L}(Y)$, we have from Theorem 6.1.1 that $\mathbf{E}[f(X)] = \mathbf{E}[f(Y)] = \int_{\mathbf{R}} f\, d\mu$.   ∎

6.2. Examples of distributions.

For a first example of a distribution of a random variable, suppose that $\mathbf{P}(X = c) = 1$, i.e. that X is always (or, at least, with probability 1) equal to some constant real number c. Then the distribution of X is the *point mass* δ_c, defined by $\delta_c(B) = \mathbf{1}_B(c)$, i.e. $\delta_c(B)$ equals 1 if $c \in B$ and equals 0 otherwise. In this case we write $X \sim \delta_c$, or $\mathcal{L}(X) = \delta_c$. From Corollary 6.1.4, since $\mathbf{P}(X = c) = 1$, we have $\mathbf{E}(X) = \mathbf{E}(c) = c$, and $\mathbf{E}(X^3 + 2) = \mathbf{E}(c^3 + 2) = c^3 + 2$, and more generally $\mathbf{E}[f(X)] = f(c)$ for any function f. In symbols, $\int_\Omega f(X(\omega)) \mathbf{P}(d\omega) \equiv \int_{\mathbf{R}} f(t)\delta_c(dt) = f(c)$. That is, the mapping $f \mapsto \mathbf{E}[f(X)]$ is an *evaluation map*.

For a second example, suppose X has the **Poisson**(5) distribution considered earlier. Then $\mathbf{P}(X \in A) = \sum_{j \in A} e^{-5} 5^j / j!$, which implies that $\mathcal{L}(X) = \sum_{j=0}^{\infty} (e^{-5} 5^j / j!)\delta_j$, a convex combination of point masses. The following proposition shows that we then have $\mathbf{E}(f(X)) = \sum_{j=0}^{\infty} f(j) e^{-5} 5^j / j!$ for any function $f : \mathbf{R} \to \mathbf{R}$.

Proposition 6.2.1.   *Suppose $\mu = \sum_i \beta_i \mu_i$, where $\{\mu_i\}$ are probability distributions, and $\{\beta_i\}$ are non-negative constants (summing to 1, if we want μ to also be a probability distribution). Then for Borel-measurable functions $f : \mathbf{R} \to \mathbf{R}$,*

$$\int f d\mu = \sum_i \beta_i \int f d\mu_i,$$

provided either side is well-defined.

Proof.   As in the proof of Theorem 6.1.1, it suffices (by linearity and the monotone convergence theorem) to check the equation when $f = \mathbf{1}_B$ is an indicator function of a Borel set B. But in this case the result follows immediately since $\mu(B) = \sum_i \beta_i \mu_i(B)$.   ∎

Clearly, any other discrete random variable can be handled similarly to the **Poisson**(5) example. Thus, discrete random variables do not present any substantial new technical issues.

For a third example of a distribution of a random variable, suppose X has the **Normal**$(0, 1)$ distribution considered earlier (henceforth denoted $N(0, 1)$). We can define its law μ_N by

$$\mu_N(B) = \int_{-\infty}^{\infty} \phi(t) \mathbf{1}_B(t) \lambda(dt), \qquad B \text{ Borel}, \qquad (6.2.2)$$

where λ is Lebesgue measure on $\mathbf{R}$ (cf. (4.4.2)) and where $\phi(t) = \frac{1}{\sqrt{2\pi}} e^{-t^2/2}$. We note that for a mathematically complete definition, it is necessary to use the Lebesgue integral rather than the Riemann integral. Indeed, the Riemann integral is undefined unless B is a rather simple set (e.g. a finite union of intervals, or more generally a set whose boundary has measure 0), while we need $\mu_N(B)$ to be defined for all Borel sets B. Furthermore, since ϕ is continuous it is Borel-measurable (Proposition 3.1.8), so Lebesgue integrals such as (6.2.2) make sense.

Similarly, given *any* Borel-measurable function (called a *density function*) f such that $f \geq 0$ and $\int_{-\infty}^{\infty} f(t)\lambda(dt) = 1$, we can define a law μ by

$$\mu(B) = \int_{-\infty}^{\infty} f(t)\mathbf{1}_B(t)\lambda(dt), \qquad B \text{ Borel}.$$

We shall sometimes write this as $\mu(B) = \int_B f$ or $\mu(B) = \int_B f(t)\,\lambda(dt)$, or even as $\mu(dt) = f(t)\,\lambda(dt)$ (where such equalities of "differentials" have the interpretation that the two sides are equal when integrated over $t \in B$ for any Borel B, i.e. that $\int_B \mu(dt) = \int_B f(t)\lambda(dt)$ for all B). We shall also write this as $\frac{d\mu}{d\lambda} = f$, and shall say that μ is absolutely continuous with respect to λ, and that f is the density for μ with respect to λ. We then have the following.

Proposition 6.2.3.   *Suppose μ has density f with respect to λ. Then for any Borel-measurable function $g : \mathbf{R} \to \mathbf{R}$,*

$$\mathbf{E}_\mu(g) \equiv \int_{-\infty}^{\infty} g(t)\mu(dt) = \int_{-\infty}^{\infty} g(t)f(t)\lambda(dt),$$

provided either side is well-defined. In words, to compute the integral of a function with respect to μ, it suffices to compute the integral of the function times the density with respect to λ.

Proof.   Once again, it suffices to check the equation when $g = \mathbf{1}_B$ is an indicator function of a Borel set B. But in that case, $\int g(t)\mu(dt) = \int \mathbf{1}_B(t)\mu(dt) = \mu(B)$, while $\int g(t)f(t)\lambda(dt) = \int \mathbf{1}_B(t)f(t)\lambda(dt) = \mu(B)$ by definition. The result follows.   ∎

By combining Theorem 4.4.1 and Proposition 6.2.3, it is possible to do explicit computations with absolutely-continuous random variables. For example, if $X \sim N(0,1)$, then

$$\mathbf{E}(X) = \int t\,\mu_N(dt) = \int t\,\phi(t)\lambda(dt) = \int_{-\infty}^{\infty} t\,\phi(t)\,dt,$$

and more generally

$$\mathbf{E}(g(X)) = \int g(t)\,\mu_N(dt) = \int g(t)\,\phi(t)\,\lambda(dt) = \int_{-\infty}^{\infty} g(t)\,\phi(t)\,dt$$

for any Riemann-integrable function g; here the last expression is an ordinary, old-fashioned, calculus-style Riemann integral. It can be computed in this manner that $\mathbf{E}(X) = 0$, $\mathbf{E}(X^2) = 1$, $\mathbf{E}(X^4) = 3$, etc.

For an example combining Propositions 6.2.3 and 6.2.1, suppose that $\mathcal{L}(X) = \frac{1}{4}\delta_1 + \frac{1}{4}\delta_2 + \frac{1}{2}\mu_N$, where μ_N is again the $N(0,1)$ distribution. Then $\mathbf{E}(X) = \frac{1}{4}(1) + \frac{1}{4}(2) + \frac{1}{2}(0) = \frac{3}{4}$, $\mathbf{E}(X^2) = \frac{1}{4}(1) + \frac{1}{4}(4) + \frac{1}{2}(1) = \frac{7}{4}$, and so on. Note, however, that it is *not* the case that $\mathbf{Var}(X)$ equals the corresponding linear combination of variances (indeed, the variance of a point-mass is 0, so that the corresponding linear combinations of variances is $\frac{1}{4}(0) + \frac{1}{4}(0) + \frac{1}{2}(1) = \frac{1}{2}$); rather, the formula $\mathbf{Var}(X) = \mathbf{E}(X^2) - \mathbf{E}(X)^2 = \frac{7}{4} - (\frac{3}{4})^2 = \frac{19}{16}$ should be used.

Exercise 6.2.4. Why does Proposition 6.2.1 not imply that $\mathbf{Var}(X)$ equals the corresponding linear combination of variances?

6.3. Exercises.

Exercise 6.3.1. Let $(\Omega, \mathcal{F}, P)$ be Lebesgue measure on $[0,1]$, and set

$$X(\omega) = \begin{cases} 1, & 0 \le \omega < 1/4 \\ 2\omega^2, & 1/4 \le \omega < 3/4 \\ \omega^2, & 3/4 \le \omega \le 1. \end{cases}$$

Compute $P(X \in A)$ where
(a) $A = [0,1]$.
(b) $A = [\frac{1}{2}, 1]$.

Exercise 6.3.2. Suppose $P(Z = 0) = P(Z = 1) = \frac{1}{2}$, that $Y \sim N(0,1)$, and that Y and Z are independent. Set $X = YZ$. What is the law of X?

Exercise 6.3.3. Let $X \sim \mathbf{Poisson}(5)$.
(a) Compute $\mathbf{E}(X)$ and $\mathbf{Var}(X)$.
(b) Compute $\mathbf{E}(3^X)$.

Exercise 6.3.4. Compute $\mathbf{E}(X)$, $\mathbf{E}(X^2)$, and $\mathbf{Var}(X)$, where the law of X is given by
(a) $\mathcal{L}(X) = \frac{1}{2}\delta_1 + \frac{1}{2}\lambda$, where λ is Lebesgue measure on $[0,1]$.
(b) $\mathcal{L}(X) = \frac{1}{3}\delta_2 + \frac{2}{3}\mu_N$, where μ_N is the standard normal distribution $N(0,1)$.

Exercise 6.3.5. Let X and Z be independent, with $X \sim N(0,1)$, and with $\mathbf{P}(Z = 1) = \mathbf{P}(Z = -1) = 1/2$. Let $Y = XZ$ (i.e., Y is the product of X and Z).

(a) Prove that $Y \sim N(0,1)$.
(b) Prove that $\mathbf{P}(|X| = |Y|) = 1$.
(c) Prove that X and Y are *not* independent.
(d) Prove that $\mathbf{Cov}(X,Y) = 0$.
(e) It is sometimes claimed that if X and Y are normally distributed random variables with $\mathbf{Cov}(X,Y) = 0$, then X and Y must be independent. Is that claim correct?

Exercise 6.3.6. Let X and Y be random variables on some probability triple $(\Omega, \mathcal{F}, \mathbf{P})$. Suppose $\mathbf{E}(X^4) < \infty$, and that $\mathbf{P}[m \le X \le z] = \mathbf{P}[m \le Y \le z]$ for all integers m and all $z \in \mathbf{R}$. Prove or disprove that we necessarily have $\mathbf{E}(Y^4) = \mathbf{E}(X^4)$.

Exercise 6.3.7. Let X be a random variable, and let $F_X(x)$ be its cumulative distribution function. For fixed $x \in \mathbf{R}$, we know by right-continuity that $\lim_{y \searrow x} F_X(y) = F_X(x)$.
(a) Give a necessary and sufficient condition that $\lim_{y \nearrow x} F_X(y) = F_X(x)$.
(b) More generally, give a formula for $F_X(x) - (\lim_{y \nearrow x} F_X(y))$, in terms of a simple property of X.

Exercise 6.3.8. Consider the statement: $f(x) = \left(f(x)\right)^2$ for all $x \in \mathbf{R}$.
(a) Prove that the statement is true for all indicator functions $f = 1_B$.
(b) Prove that the statement is *not* true for the identity function $f(x) = x$.
(c) Why does this fact not contradict the method of proof of Theorem 6.1.1?

6.4. Section summary.

This section defined the distribution (or law), $\mathcal{L}(X)$, of a random variable X, to be a corresponding distribution on the real line. It proved that $\mathcal{L}(X)$ is completely determined by the cumulative distribution function, $F_X(x) = \mathbf{P}(X \le x)$, of X. It proved that expectation $\mathbf{E}(f(X))$ of any function of X can be computed (in principle) once $\mathcal{L}(X)$ or $F_X(x)$ is known. It then considered a number of examples of distributions of random variables, including discrete and continuous random variables and various combinations of them. It provided a number of results for computing expected values with respect to such distributions.

7. Stochastic processes and gambling games.

Now that we have covered most of the essential foundations of rigorous probability theory, it is time to get "moving", i.e. to consider random *processes* rather than just static random variables.

A (discrete time) *stochastic process* is simply a sequence $X_0, X_1, X_2, \ldots$ of random variables defined on some fixed probability triple $(\Omega, \mathcal{F}, \mathbf{P})$. The random variables $\{X_n\}$ are typically *not* independent. In this context we often think of n as representing time; thus, X_n represents the value of a random quantity at the time n.

For a specific example, let $(r_1, r_2, \ldots)$ be the result of infinite fair coin tossing (so that $\{r_i\}$ are independent, and each r_i equals 0 or 1 with probability $\frac{1}{2}$; see Subsection 2.6), and set

$$X_0 = 0; \quad X_n = r_1 + r_2 + \ldots + r_n, \ n \geq 1; \qquad (7.0.1)$$

thus, X_n represents the number of heads obtained up to time n. Alternatively, we might set

$$X_0 = 0; \quad X_n = 2(r_1 + r_2 + \ldots + r_n) - n, \ n \geq 1; \qquad (7.0.2)$$

then X_n represents the number of heads *minus* the number of tails obtained up to time n. This last example suggests a gambling game: each time we obtain a head we increase X_n by 1 (i.e., we "win"), while each time we obtain a tail we decrease X_n by 1 (i.e., we "lose").

To allow for non-fair games, we might wish to generalise (7.0.2) to

$$X_0 = 0; \quad X_n = Z_1 + Z_2 + \ldots + Z_n, \ n \geq 1,$$

where the $\{Z_i\}$ are assumed to be i.i.d. random variables, satisfying $\mathbf{P}(Z_i = 1) = p$ and $\mathbf{P}(Z_i = -1) = 1 - p$, for some fixed $0 < p < 1$. (If $p = \frac{1}{2}$ then this is equivalent to (7.0.2), with $Z_i = 2r_i - 1$.)

This raises an immediate issue: can we be sure that such random variables $\{Z_i\}$ even exist? In fact the answer to this question is yes, as the following subsection shows.

7.1. A first existence theorem.

We here show the existence of sequences of independent random variables having prescribed distributions.

Theorem 7.1.1.   Let $\mu_1, \mu_2, \ldots$ be any sequence of Borel probability measures on $\mathbf{R}$. Then there exists a probability space $(\Omega, \mathcal{F}, \mathbf{P})$, and random variables $X_1, X_2, \ldots$ defined on $(\Omega, \mathcal{F}, \mathbf{P})$, such that $\{X_n\}$ are independent, and $\mathcal{L}(X_n) = \mu_n$.

We begin with a lemma.

Lemma 7.1.2. *Let U be a random variable whose distribution is Lebesgue measure (i.e., the uniform distribution) on $[0, 1]$. Let F be any cumulative distribution function, and set $\phi(u) = \inf\{x; F(x) \geq u\}$ for $0 < u < 1$. Then $\mathbf{P}(\phi(U) \leq x) = F(x)$ for each $x \in \mathbf{R}$; in words, the cumulative distribution function of $\phi(U)$ is F.*

Proof. Since F is right-continuous, we have that $\inf\{x; F(x) \geq u\} = \min\{x; F(x) \geq u\}$, i.e. the infimum is actually obtained. It follows that $\phi(u) \leq x$ if and only if $u \leq F(x)$. Hence, since $0 \leq F(x) \leq 1$, we obtain that $\mathbf{P}(\phi(U) \leq x) = \mathbf{P}(U \leq F(x)) = F(x)$. ∎

Proof of Theorem 7.1.1. We let $(\Omega, \mathcal{F}, \mathbf{P})$ be infinite independent fair coin tossing, so that $r_1, r_2, \ldots$ are i.i.d. with $\mathbf{P}(r_i = 0) = \mathbf{P}(r_i = 1) = \frac{1}{2}$. Let $\{Z_{ij}\}$ be a two-dimensional array filled by these r_i, as follows:

$$\begin{pmatrix} Z_{11} & Z_{12} & Z_{13} & \cdots \\ Z_{21} & Z_{22} & Z_{23} & \cdots \\ Z_{31} & Z_{32} & Z_{33} & \cdots \\ Z_{41} & Z_{42} & Z_{43} & \cdots \\ \vdots & \vdots & \vdots & \end{pmatrix} \equiv \begin{pmatrix} r_1 & r_3 & r_6 & \cdots \\ r_2 & r_5 & \cdots & \\ r_4 & r_8 & \cdots & \\ r_7 & \cdots & & \\ \vdots & \vdots & & \end{pmatrix}.$$

Hence, $\{Z_{ij}\}$ are independent, with $\mathbf{P}(Z_{ij} = 0) = \mathbf{P}(Z_{ij} = 1) = \frac{1}{2}$.

Then, for each $n \in \mathbf{N}$, we set $U_n = \sum_{k=1}^{\infty} Z_{nk}/2^k$. By Corollary 3.5.3, the $\{U_n\}$ are independent. Furthermore, by the way the U_n were constructed, we have $\mathbf{P}\left(\frac{j}{2^k} \leq U_n < \frac{j+1}{2^k}\right) = \frac{1}{2^k}$ for $k \in \mathbf{N}$ and $0 \leq j < 2^k$. By additivity and continuity of probabilities, this implies that $\mathbf{P}(a \leq U_n < b) = b - a$ whenever $0 \leq a < b \leq 1$. Hence, by Proposition 2.5.8, each U_n follows the uniform distribution (i.e. Lebesgue measure) on $[0, 1]$.

Finally, we set $F_n(x) = \mu_n((-\infty, x])$ for $x \in \mathbf{R}$, set $\phi_n(u) = \inf\{x; u \leq F_n(x)\}$ for $0 < u < 1$, and set $X_n = \phi_n(U_n)$. Then $\{X_n\}$ are independent by Proposition 3.2.3, and $X_n \sim \mu_n$ by Lemma 7.1.2, as required. ∎

7.2. Gambling and gambler's ruin.

By Theorem 7.1.1, for fixed $0 < p < 1$ we can find random variables $\{Z_i\}$ which are i.i.d. with $\mathbf{P}(Z_i = 1) = p$ and $\mathbf{P}(Z_i = -1) = 1 - p \equiv q$. We then set $X_n = a + Z_1 + Z_2 + \ldots + Z_n$ (with $X_0 \equiv a$) for some fixed integer a. We shall interpret X_n as a gambling player's "fortune" (in dollars) at time n when repeatedly making \$1 bets, and shall refer to the stochastic process $\{X_n\}$ as *simple random walk*. Thus, our player begins with \$$a$, and at each time has probability p of winning \$1 and probability $q = 1 - p$ of losing \$1.

We first note the distribution of X_n. Indeed, clearly $\mathbf{P}(X_n = a + k) = 0$ unless $-n \le k \le n$ with $n + k$ even. For such k, there are $\binom{n}{\frac{n+k}{2}}$ different possible sequences $Z_1, \ldots, Z_n$ such that $X_n = a + k$, namely all sequences consisting of $\frac{n+k}{2}$ symbols $+1$ and $\frac{n-k}{2}$ symbols -1. Furthermore, each such sequence has probability $p^{\frac{n+k}{2}} q^{\frac{n-k}{2}}$. We conclude that

$$\mathbf{P}(X_n = a + k) = \binom{n}{\frac{n+k}{2}} p^{\frac{n+k}{2}} q^{\frac{n-k}{2}}, \qquad -n \le k \le n, \quad n + k \text{ even},$$

with $\mathbf{P}(X_n = a + k) = 0$ otherwise.

This is a rather "static" observation about the process $\{X_n\}$; of greater interest are questions which depend on its time-evolution. One such question is the *gambler's ruin* problem, defined as follows. Suppose that $0 < a < c$, and let $\tau_0 = \inf\{n \ge 0; X_n = 0\}$ and $\tau_c = \inf\{n \ge 0; X_n = c\}$ be the *first hitting time* of 0 and c, respectively. (These infima are taken to be $+\infty$ if the condition is satisfied for no n.) The gambler's ruin question is, what is $\mathbf{P}(\tau_c < \tau_0)$? That is, what is the probability that the player's fortune will reach the value c before it reaches the value 0. Informally, what is the probability that the gambler gets rich before going broke? (Note that $\{\tau_c < \tau_0\}$ includes the case when $\tau_0 = \infty$ while $\tau_c < \infty$; but it does *not* include the case when $\tau_c = \tau_0 = \infty$.)

Solving this question is not straightforward, since there is no limit to how long it will take until either the fortune c or the fortune 0 is reached. However, by using the right trick, the solution presents itself. We set $s(a) = s_{c,p}(a) = \mathbf{P}(\tau_c < \tau_0)$. Writing the dependence on a explicitly will allow us to *vary* a, and to relate $s(a)$ to $s(a-1)$ and $s(a+1)$. Indeed, for $1 \le a \le c-1$ we have that

$$\begin{aligned} s(a) &= \mathbf{P}(\tau_c < \tau_0) \\ &= \mathbf{P}(Z_1 = -1, \ \tau_c < \tau_0) + \mathbf{P}(Z_1 = +1, \ \tau_c < \tau_0) \qquad (7.2.1) \\ &= q\, s(a-1) + p\, s(a+1). \end{aligned}$$

That is, $s(a)$ is a simple convex combination of $s(a-1)$ and $s(a+1)$; this is the key. We further have by definition that $s(0) = 0$ and $s(c) = 1$.

Now, (7.2.1) gives $c-1$ equations, for the $c-1$ unknowns $s(1), \ldots, s(c-1)$. This system of equations can then be solved in several different ways (see Exercises 7.2.4, 7.2.5, and 7.2.6 below), to obtain that

$$s(a) = s_{c,p}(a) = \frac{1 - \left(\frac{q}{p}\right)^a}{1 - \left(\frac{q}{p}\right)^c}, \qquad p \neq \frac{1}{2}. \tag{7.2.2}$$

and

$$s(a) = s_{c,p}(a) = a/c, \qquad p = \frac{1}{2}. \tag{7.2.3}$$

(This last equation is suggestive: for a fair game ($p = \frac{1}{2}$), with probability a/c you end up with c dollars; and the product of these two quantities is your initial fortune a. We will consider this issue again when we study martingales; see in particular Exercises 14.4.10 and 14.4.11.)

Exercise 7.2.4. Verify that (7.2.2) and (7.2.3) satisfy (7.2.1), and also satisfy $s(0) = 0$ and $s(c) = 1$.

Exercise 7.2.5. Solve equation (7.2.1) by direct algebra, as follows.
(a) Show that (7.2.1) implies that for $1 \leq a \leq c - 1$, $s(a + 1) - s(a) = \frac{q}{p}(s(a) - s(a - 1))$.
(b) Show that this implies that for $0 \leq a \leq c - 1$, $s(a + 1) - s(a) = \left(\frac{q}{p}\right)^a s(1)$.
(c) Show that this implies that for $0 \leq a \leq c$, $s(a) = \sum_{i=0}^{a-1} \left(\frac{q}{p}\right)^i s(1)$.
(d) Solve for $s(1)$, and verify (7.2.2) and (7.2.3).

Exercise 7.2.6. Solve equation (7.2.1) using the theory of *difference equations*, as follows.
(a) Show that the corresponding "characteristic equation" $t^0 = qt^{-1} + pt^1$ has two distinct roots t_1 and t_2 when $p \neq 1/2$, and one double root t_3 when $p = 1/2$. Solve for t_1, t_2, and t_3.
(b) When $p \neq 1/2$, the theory of difference equations says that we must have $s_{c,p}(a) = C_1(t_1)^a + C_2(t_2)^a$ for some constants C_1 and C_2. Assuming this, use the *boundary conditions* $s_{c,p}(0) = 0$ and $s_{c,p}(c) = 1$ to solve for C_1 and C_2. Verify (7.2.2).
(c) When $p = 1/2$, the theory of difference equations says that we must have $s_{c,p}(a) = C_3(t_3)^a + C_4 a(t_3)^a$ for some constants C_3 and C_4. Assuming this, use the boundary conditions to solve for C_3 and C_4. Verify (7.2.3).

As a specific example, suppose you start with \$9,700 (i.e., $a = 9700$) and your goal is to win \$10,000 before going broke (i.e., $c = 10000$). If $p = \frac{1}{2}$, then your probability of success is $a/c = 0.97$, which is very high; on the other hand, if $p = 0.49$, then your probability of success is given by

$$\left[1 - \left(\frac{0.51}{0.49}\right)^{9700}\right] \bigg/ \left[1 - \left(\frac{0.51}{0.49}\right)^{10000}\right],$$

which is approximately 6.1×10^{-6}, or about one chance in 163,000. This shows rather dramatically that even a small disadvantage on each bet can lead to a very large disadvantage in the long run!

Now let $r(a) = r_{c,p}(a) = \mathbf{P}(\tau_0 < \tau_c)$ be the probability that our gambler goes broke before reaching the desired fortune. Then clearly $s(a)$ and $r(a)$ are related. Indeed, by "considering the bank's point of view", we see immediately that $r_{c,p}(a) = s_{c,1-p}(c-a)$ (that is, the chance of going broke before obtaining c dollars, when starting with a dollars and having probability p of winning each bet, is the same as the chance of obtaining c dollars before going broke, when starting with $c - a$ dollars and having probability $1 - p$ of winning each bet), so that

$$r_{c,p}(a) = \begin{cases} \frac{1-(\frac{p}{q})^{c-a}}{1-(\frac{p}{q})^{c}}, & p \neq \frac{1}{2} \\ \\ \frac{c-a}{c}, & p = \frac{1}{2}. \end{cases}$$

Finally, let us consider the probability that the gambler will *eventually* go broke if they never stop gambling, i.e. $\mathbf{P}(\tau_0 < \infty)$ without regard to any target fortune c. Well, if we let $H_c = \{\tau_0 < \tau_c\}$, then clearly $\{H_c\}$ is *increasing* up to $\{\tau_0 < \infty\}$, as $c \to \infty$. Hence, by continuity of probabilities,

$$\mathbf{P}(\tau_0 < \infty) = \lim_{c \to \infty} \mathbf{P}(H_c) = \lim_{c \to \infty} r_{c,p}(a) = \begin{cases} 1, & p \leq \frac{1}{2} \\ \\ \left(\frac{q}{p}\right)^{a}, & p > \frac{1}{2}. \end{cases} \quad (7.2.7)$$

Thus, if $p \leq \frac{1}{2}$ (i.e., if the gambler has no advantage), then the gambler is certain to eventually go broke. On the other hand, if say $p = 0.6$ and $a = 1$, then the gambler has probability $1/3$ of never going broke.

7.3. Gambling policies.

Suppose now that our gambler is allowed to *choose* how much to bet each time. That is, the gambler can choose random variables W_n so that their fortune at time n is given by

$$X_n = a + W_1 Z_1 + W_2 Z_2 + \ldots + W_n Z_n,$$

with $\{Z_n\}$ as before. To avoid "cheating", we shall insist that $W_n \geq 0$ (you can't bet a negative amount), and also that $W_n = f_n(Z_1, Z_2, \ldots, Z_{n-1})$ is a function only of the *previous* bet results (you can't know the result of a bet before you choose how much to wager). Here the $f_n : \{-1, 1\}^{n-1} \to \mathbf{R}^{\geq 0}$ are fixed deterministic functions, collectively referred to as the *gambling policy*. (If $W_n \equiv 1$ for each n, then this is equivalent to our previous gambling model.)

Note that $X_n = X_{n-1} + W_n Z_n$, with W_n and Z_n independent, and with $\mathbf{E}(Z_n) = p - q$. Hence, $\mathbf{E}(X_n) = \mathbf{E}(X_{n-1}) + (p - q)\mathbf{E}(W_n)$. Furthermore, since $W_n \geq 0$, therefore $\mathbf{E}(W_n) \geq 0$. It follows that

(a) if $p = \frac{1}{2}$, then $\mathbf{E}(X_n) = \mathbf{E}(X_{n-1}) = \ldots = \mathbf{E}(X_0) = a$, so that $\lim \mathbf{E}(X_n) = a$;

(b) if $p \leq \frac{1}{2}$, then $\mathbf{E}(X_n) \leq \mathbf{E}(X_{n-1}) \leq \ldots \leq \mathbf{E}(X_0) = a$, so that $\lim \mathbf{E}(X_n) \leq a$;

(c) if $p \geq \frac{1}{2}$, then $\mathbf{E}(X_n) \geq \mathbf{E}(X_{n-1}) \geq \ldots \geq \mathbf{E}(X_0) = a$, so that $\lim \mathbf{E}(X_n) \geq a$.

This seems simple enough, and corresponds to our intuition: if $p \leq \frac{1}{2}$ then the player's expected value can only decrease. End of story?

Perhaps not. Consider the "double 'til you win" policy, defined by

$$W_1 \equiv 1; \qquad W_n = \begin{cases} 2^{n-1}, & Z_1 = Z_2 = \ldots = Z_{n-1} = -1 \\ 0, & \text{otherwise} \end{cases}, \qquad n \geq 2.$$

That is, we first bet \$1. Each time we lose, we *double* our bet on the succeeding turn. As soon as we win once, we bet zero from then on.

It is easily seen that, with this gambling policy, we will be up \$1 as soon as we win a bet. That is, letting $\tau = \inf\{n \geq 1; Z_n = +1\}$, we have that $X_n = a + 1$ provided that $\tau \leq n$. Now, clearly $\mathbf{P}(\tau > n) = (1 - p)^n$. Hence, assuming $p > 0$, we see that $\mathbf{P}(\tau < \infty) = 1$. It follows that $\mathbf{P}(\lim X_n = a + 1) = 1$, so that $\mathbf{E}(\lim X_n) = a + 1$. In words, with probability one we will gain \$1 with this gambling policy, for *any* positive value of p, and thus "cheat fate". How can this be, in light of (a) and (b) above?

The answer, of course, is that in this case $\mathbf{E}(\lim X_n)$ (which equals $a + 1$) is not the same as $\lim \mathbf{E}(X_n)$ (which must be $\leq a$). This is what allows us to "cheat fate". On the other hand, we may need to lose an arbitrarily large amount of money before we win our \$1, so "infinite capital" is required to follow this gambling policy. We show now that, if the fortunes X_n must remain *bounded* (i.e., if we only have a *finite* amount of capital to draw on), then $\mathbf{E}(\lim X_n)$ must indeed be the same as $\lim \mathbf{E}(X_n)$.

Theorem 7.3.1.   (*The bounded convergence theorem.*)   Let $\{X_n\}$ be a sequence of random variables, with $\lim X_n = X$. Suppose there is $K \in \mathbf{R}$ such that $|X_n| \leq K$ for all $n \in \mathbf{N}$ (i.e., the $\{X_n\}$ are uniformly bounded). Then $\mathbf{E}(X) = \lim \mathbf{E}(X_n)$.

Proof. We have from the triangle inequality that

$$|\mathbf{E}(X) - \mathbf{E}(X_n)| = |\mathbf{E}(X - X_n)| \leq \mathbf{E}(|X - X_n|).$$

We shall show that this last expression goes to 0 as $n \to \infty$. Indeed, fix $\epsilon > 0$, and set $A_n = \{\omega \in \Omega; \ |X(\omega) - X_n(\omega)| > \epsilon\}$. Then $|X(\omega) - X_n(\omega)| \leq \epsilon + 2K\mathbf{1}_{A_n}(\omega)$, so that $\mathbf{E}(|X - X_n|) \leq \epsilon + 2K\mathbf{P}(A_n)$. Hence, using Proposition 3.4.1, we have that

$$\begin{aligned}
\limsup \mathbf{E}(|X - X_n|) &\leq \ \epsilon + 2K \limsup \mathbf{P}(A_n) \\
&\leq \ \epsilon + 2K\,\mathbf{P}\left(\limsup A_n\right) \\
&= \ \epsilon,
\end{aligned}$$

since $|X(\omega) - X_n(\omega)| \to 0$ for all $\omega \in \Omega$, so that $\limsup A_n$ is the empty set. Hence, $\mathbf{E}(|X - X_n|) \to 0$, as claimed. ∎

It follows immediately that, if we are gambling with $p \leq \frac{1}{2}$, and we use any gambling policy which leaves the fortunes $\{X_n\}$ uniformly bounded, then $\lim X_n$ (if it exists) will have expected value equal to $\lim \mathbf{E}(X_n)$, and therefore be $\leq a$. So, if we have only finite capital (no matter how large), then it is not possible to cheat fate.

Finally, we consider the following question. Suppose $p < \frac{1}{2}$, and $0 < a < c$. What gambling system maximises $\mathbf{P}(\tau_c < \tau_0)$, i.e. maximises the probability that we reach the fortune c before losing all our money?

For example, suppose again that $p = 0.49$, and that $a = 9700$, $c = 10000$. If $W_n \equiv 1$ (i.e. we bet exactly \$1 each time), then we already know that $\mathbf{P}(\tau_c < \tau_0) = \left[(0.51/0.49)^{9700} - 1\right] / \left[(0.51/0.49)^{10000} - 1\right] \doteq 6.1 \times 10^{-6}$. On the other hand, if $W_n \equiv 2$, then this is equivalent to instead having $a = 4850$ and $c = 5000$, so we see that $\mathbf{P}(\tau_c < \tau_0) = \left[(0.51/0.49)^{4850} - 1\right] / \left[(0.51/0.49)^{5000} - 1\right] \doteq 2.5 \times 10^{-3}$, which is about 400 times better. In fact, if $W_n \equiv 100$, then $\mathbf{P}(\tau_c < \tau_0) = \left[(0.51/0.49)^{97} - 1\right] / \left[(0.51/0.49)^{100} - 1\right] \doteq 0.885$, which is a very favourable probability.

This example suggests that, if $p < \frac{1}{2}$ and we wish to maximise $\mathbf{P}(\tau_c < \tau_0)$, then it is best to bet in *larger* amounts, i.e. to get the game over with as quickly as possible. This is indeed true. More precisely, we define *bold play* to be the gambling strategy $W_n = \min(X_{n-1}, c - X_{n-1})$, i.e. the strategy of betting as much as possible each time. It is then a theorem that, when $p < \frac{1}{2}$, this is the optimal strategy in the sense of maximising $\mathbf{P}(\tau_c < \tau_0)$. For a proof, see e.g. Billingsley (1995, Theorem 7.3).

Disclaimer. We note that for $p \leq \frac{1}{2}$, even though to maximise $\mathbf{P}(\tau_c < \tau_0)$ it is best to bet large amounts, still overall it is best not to gamble at all. Indeed, by (7.2.7), if $p \leq \frac{1}{2}$ and you have any finite amount of money, then if you keep betting you will eventually lose it all and go broke. This is the reason few probabilists attend gambling casinos!

7.4. Exercises.

Exercise 7.4.1.   For the stochastic process $\{X_n\}$ given by (7.0.1), compute (for $n, k > 0$)
(a) $\mathbf{P}(X_n = k)$.
(b) $\mathbf{P}(\tau_k = n)$.
[Hint: These two questions do not have the same answer.]

Exercise 7.4.2.   For the stochastic process $\{X_n\}$ given by (7.0.2), compute (for $n, k > 0$)
(a) $\mathbf{P}(X_n = k)$.
(b) $\mathbf{P}(X_n > 0)$.

Exercise 7.4.3.   Prove that there exist random variables Y and Z such that $\mathbf{P}(Y = 1) = \mathbf{P}(Y = -1) = \mathbf{P}(Z = 1) = \mathbf{P}(Z = -1) = \frac{1}{4}$, $\mathbf{P}(Y = 0) = \mathbf{P}(Z = 0) = \frac{1}{2}$, and such that $\mathbf{Cov}(Y, Z) = \frac{1}{4}$. (In particular, Y and Z are *not* independent.) [Hint: First use Theorem 7.1.1 to construct independent random variables X_1, X_2, and X_3 each having certain two-point distributions. Then construct Y and Z as functions of X_1, X_2, and X_3.]

Exercise 7.4.4.   For the gambler's ruin model of Subsection 7.2, with $c = 10000$ and $p = 0.49$, find the smallest positive integer a such that $s_{c,p}(a) \geq \frac{1}{2}$. Interpret your result in plain English.

Exercise 7.4.5.   For the gambler's ruin model of Subsection 7.2, let $\beta_n = \mathbf{P}(\min(\tau_0, \tau_c) > n)$ be the probability that the player's fortune has not hit 0 or c by time n.
(a) Find any explicit, simple expression γ_n such that $\beta_n < \gamma_n$ for all $n \in \mathbf{N}$, and such that $\lim_{n \to \infty} \gamma_n = 0$.
(b) Find any explicit, simple expression α_n such that $\beta_n > \alpha_n > 0$ for all $n \in \mathbf{N}$.

Exercise 7.4.6.   Let $\{W_n\}$ be i.i.d. with $\mathbf{P}[W_n = +1] = \mathbf{P}[W_n = 0] = 1/4$ and $\mathbf{P}[W_n = -1] = 1/2$, and let a be a positive integer. Let $X_n = a + W_1 + \ldots + W_n$, and let $\tau_0 = \inf\{n \geq 0; X_n = 0\}$. Compute $\mathbf{P}(\tau_0 < \infty)$. [Hint: Let $\{Y_n\}$ be like $\{X_n\}$, except with immediate repetitions of values omitted. Is $\{Y_n\}$ a simple random walk? With what parameter p?]

Exercise 7.4.7. Verify explicitly that $r_{c,p}(a) + s_{c,p}(a) = 1$ for all a, c, and p.

Exercise 7.4.8. In gambler's ruin, recall that $\{\tau_c < \tau_0\}$ is the event that the player eventually wins, and $\{\tau_0 < \tau_c\}$ is the event that the player eventually loses.
(a) Give a similar plain-English description of the complement of the union of these two events, i.e. $(\{\tau_c < \tau_0\} \cup \{\tau_0 < \tau_c\})^C$.
(b) Give *three* different proofs that the event described in part (a) has probability 0: one using Exercise 7.4.7; a second using Exercise 7.4.5; and a third recalling how the probabilities $s_{c,p}(a)$ were computed in the text, and seeing to what extent the computation would have differed if we had instead replaced $s_{c,p}(a)$ by $S_{c,p}(a) = \mathbf{P}(\tau_c \leq \tau_0)$.
(c) Prove that, if $c \geq 4$, then the event described in part (a) contains uncountably many outcomes (i.e., that uncountably many different sequences $Z_1, Z_2, \ldots$ correspond to this event, even though it has probability 0). [Hint: This is not entirely dissimilar from the analysis of the *Cantor set* in Subsection 2.4.]

Exercise 7.4.9. For the gambling policies model of Subsection 7.3, consider the "triple 'til you win" policy defined by $W_1 \equiv 1$, and for $n \geq 2$, $W_n = 3^{n-1}$ if $Z_1 = \ldots = Z_{n-1} = -1$ otherwise $W_n = 0$.
(a) Prove that, with probability 1, the limit $\lim_{n\to\infty} X_n$ exists.
(b) Describe precisely the distribution of $\lim_{n\to\infty} X_n$.

Exercise 7.4.10. Consider the gambling policies model, with $p = 1/3$, $a = 6$, and $c = 8$.
(a) Compute the probability $s_{c,p}(a)$ that the player will win (i.e. hit c before hitting 0) if they bet \$1 each time (i.e. if $W_n \equiv 1$).
(b) Compute the probability that the player will win if they bet \$2 each time (i.e. if $W_n \equiv 2$).
(c) Compute the probability that the player will win if they employ the strategy of Bold Play (i.e., if $W_n = \min(X_{n-1}, c - X_{n-1})$). [Hint: While it is difficult to do explicit computations involving Bold Play in general, here the numbers are small enough that it is not difficult.]

7.5. Section summary.

This section introduced the concept of stochastic processes, from the point of view of gambling games. It proved a general theorem about the existence of sequences of independent random variables having arbitrary specified distributions. It used this to define several models of gambling games. If the player bets \$1 each time, and has independent probability p of winning each bet, then explicit formulae were developed for the probability that the player achieves some specified target fortune c before losing all their money. The formula is very sensitive to values of p near $\frac{1}{2}$.

If the player is allowed to choose how much to bet at each stage, then with clever betting they can "cheat fate" and always win. However, this requires them to have infinite capital available. If their capital is finite, and $p \leq \frac{1}{2}$, then on average they will always lose money by the Bounded Convergence Theorem.

8. Discrete Markov chains.

In this section, we consider the general notion of a (discrete time, discrete space, time-homogeneous) Markov chain.

A *Markov chain* is characterised by three ingredients: a *state space* S which is a finite or countable set; an *initial distribution* $\{\nu_i\}_{i \in S}$ consisting of non-negative numbers summing to 1; and *transition probabilities* $\{p_{ij}\}_{i,j \in S}$ consisting of non-negative numbers with $\sum_{j \in S} p_{ij} = 1$ for each $i \in S$.

Intuitively, a Markov chain represents the random motion of some particle moving around the space S. ν_i represents the probability that the particle starts at the point i, while p_{ij} represents the probability that, *if* the particle is at the point i, it will then jump to the point j on the next step.

More formally, a Markov chain is defined to be a sequence of random variables $X_0, X_1, X_2, \ldots$ taking values in the set S, such that

$$\mathbf{P}(X_0 = i_0, X_1 = i_1, \ldots, X_n = i_n) = \nu_{i_0} p_{i_0 i_1} p_{i_1 i_2} \cdots p_{i_{n-1} i_n}$$

for any $n \in \mathbf{N}$ and any choice of $i_0, i_1, \ldots, i_n \in S$. (Note that, for these $\{X_n\}$ to be random variables in the sense of Definition 3.1.1, we need to have $S \subseteq \mathbf{R}$; however, if we allow more general random variables as in Remark 3.1.10, then this restriction is not necessary.) It then follows that, for example,

$$\mathbf{P}(X_1 = j) = \sum_{i \in S} \mathbf{P}(X_0 = i,\ X_1 = j) = \sum_{i \in S} \nu_i p_{ij}.$$

This also has a matrix interpretation: writing $[\mathbf{p}]$ for the matrix $\{p_{ij}\}_{i,j \in S}$, and $[\mu^{(n)}]$ for the row-vector $\{\mathbf{P}(X_n = i)\}_{i \in S}$, we have $[\mu^{(1)}] = [\mu^{(0)}]\,[\mathbf{p}]$, and more generally $[\mu^{(n+1)}] = [\mu^{(n)}]\,[\mathbf{p}]$.

We present some simple examples of Markov chains here. Note that, except in the first example, we do not bother to specify initial probabilities $\{\nu_i\}$; we shall see that initial probabilities are often not crucial when studying a chain's properties.

Example 8.0.1. (Simple random walk.) Let $S = \mathbf{Z}$ be the set of all integers. Fix $a \in \mathbf{Z}$, and let $\nu_a = 1$, with $\nu_i = 0$ for $i \neq a$. Fix a real number p with $0 < p < 1$, and let $p_{i,i+1} = p$ and $p_{i,i-1} = 1 - p$ for each $i \in \mathbf{Z}$, with $p_{ij} = 0$ if $j \neq i \pm 1$. Thus, this Markov chain begins at the point a (with probability 1), and at each step either increases by 1 (with probability p) or decreases by 1 (with probability $1-p$). It is easily seen that this Markov chain corresponds precisely to the gambling game of Subsection 7.2.

Example 8.0.2. Let $S = \{1, 2, 3\}$ consist of just three elements, and define the transition probabilities $\{p_{ij}\}$ in matrix form by

$$(p_{ij}) = \begin{pmatrix} 0 & 1/2 & 1/2 \\ 1/3 & 1/3 & 1/3 \\ 1/4 & 1/4 & 1/2 \end{pmatrix}$$

(so that $p_{31} = \frac{1}{4}$, etc.). This Markov chain jumps around on the three points $\{1, 2, 3\}$ in a random and interesting way.

Example 8.0.3. (Random walk on $\mathbf{Z}/(d)$.) Let $S = \{0, 1, 2, \ldots, d-1\}$, and define the transition probabilities by

$$p_{ij} = \begin{cases} \frac{1}{3}, & i = j \text{ or } i = j+1 (\bmod d) \text{ or } i = j-1 (\bmod d); \\ 0, & \text{otherwise}. \end{cases}$$

If we think of the d elements of S as arranged in a circle, then our particle, at each step, either stays where it is, or moves one step clockwise, or moves one step counter-clockwise, each with probability $\frac{1}{3}$.

Example 8.0.4. (Ehrenfest's urn.) Consider two urns, Urn 1 and Urn 2. Suppose there are d balls divided between the two urns. Suppose at each step, we choose one ball uniformly at random from among the d balls, and switch it to the opposite urn. We let X_n be the number of balls in Urn 1 at time n. Thus, $S = \{0, 1, 2, \ldots, d\}$, with $p_{i,i-1} = i/d$ and $p_{i,i+1} = (d-i)/d$, for $0 \le i \le d$ (with $p_{ij} = 0$ if $j \ne i \pm 1$). Thus, this Markov chain moves randomly among the possible numbers $\{0, 1, \ldots, d\}$ of balls in Urn 1 at each time. One might expect that, if d is large and the Markov chain is run for a long time, that there would most likely be approximately $d/2$ balls in Urn 1. (We shall consider such questions in Subsection 8.3 below.)

We note that we can also interpret Markov chains in terms of conditional probability. Recall that, if A and B are events with $\mathbf{P}(B) > 0$, then the *conditional probability* of A given B is $\mathbf{P}(A|B) = \frac{\mathbf{P}(A \cap B)}{\mathbf{P}(B)}$; intuitively, it is the probabilistic proportion of the event B which is also contained in the event A. Thus, for a Markov chain, if $\mathbf{P}(X_k = i) > 0$, then

$$\begin{aligned} \mathbf{P}(X_{k+1} = j \,|\, X_k = i) &= \frac{\mathbf{P}(X_k=i,\, X_{k+1}=j)}{\mathbf{P}(X_k=i)} \\ &= \frac{\sum_{i_0, i_1, \ldots, i_{k-1}} \nu_{i_0} p_{i_0 i_1} p_{i_1 i_2} \cdots p_{i_{k-2} i_{k-1}} p_{i_{k-1} i} p_{ij}}{\sum_{i_0, i_1, \ldots, i_{k-1}} \nu_{i_0} p_{i_0 i_1} p_{i_1 i_2} \cdots p_{i_{k-2} i_{k-1}} p_{i_{k-1} i}} \\ &= p_{ij}. \end{aligned}$$

This formally justifies the notion of p_{ij} as a transition probability. Note also that this conditional probability does not depend on the starting time

k, justifying the notion of time homogeneity. (The only reason we do not take this conditional probability as our definition of a Markov chain, is that the conditional probability is not well-defined if $\mathbf{P}(X_k = i) = 0$.)

We can similarly compute, for any $n \in \mathbf{N}$, again assuming $\mathbf{P}(X_k = i) > 0$, that $\mathbf{P}(X_{k+n} = j \,|\, X_k = i)$ is equal to

$$\sum_{i_{k+1}, i_{k+2}, \ldots, i_{k+n-1}} p_{i i_{k+1}} p_{i_{k+1} i_{k+2}} \cdots p_{i_{k+n-2} i_{k+n-1}} p_{i_{k+n-1} j} \equiv p_{ij}^{(n)} ;$$

here $p_{ij}^{(n)}$ is an nth *order transition probability*. Note again that this probability does not depend on the starting time k (despite appearances to the contrary). By convention, we set

$$p_{ij}^{(0)} = \delta_{ij} = \left\{ \begin{array}{ll} 1, & i = j \\ 0, & \text{otherwise} \end{array} \right. , \qquad (8.0.5)$$

i.e. in zero time units the Markov chain just stays where it is.

8.1. A Markov chain existence theorem.

To rigorously study Markov chains, we need to be sure that they exist. Fortunately, this is relatively straightforward.

Theorem 8.1.1. *Given a non-empty countable set S, and non-negative numbers $\{\nu_i\}_{i \in S}$ and $\{p_{ij}\}_{i,j \in S}$, with $\sum_j \nu_j = 1$ and $\sum_j p_{ij} = 1$ for each $i \in S$, there exists a probability triple $(\Omega, \mathcal{F}, \mathbf{P})$, and random variables $X_0, X_1, \ldots$ defined on $(\Omega, \mathcal{F}, \mathbf{P})$, such that*

$$\mathbf{P}(X_0 = i_0, X_1 = i_1, \ldots, X_n = i_n) = n u_{i_0} p_{i_0 i_1} \cdots p_{i_{n-1} i_n}$$

for all $n \in \mathbf{N}$ and all $i_0, \ldots, i_n \in S$.

Proof. We let $(\Omega, \mathcal{F}, \mathbf{P})$ be Lebesgue measure on $[0, 1]$. We construct the random variables $\{X_n\}$ as follows.

1. Partition $[0, 1]$ into intervals $\{I_i^{(0)}\}_{i \in S}$, with $\text{length}(I_i^{(0)}) = \nu_i$.
2. Partition each $I_i^{(0)}$ into intervals $\{I_{ij}^{(1)}\}_{i,j \in S}$, with $\text{length}(I_{ij}^{(1)}) = \nu_i p_{ij}$.
3. Inductively partition $[0, 1]$ into intervals $\{I_{i_0 i_1 \ldots i_n}^{(n)}\}_{i_0, i_1, \ldots, i_n \in S}$, such that $I_{i_0 i_1 \ldots i_n}^{(n)} \subseteq I_{i_0 i_1 \ldots i_{n-1}}^{(n-1)}$ and $\text{length}(I_{i_0 i_1 \ldots i_n}^{(n)}) = \nu_{i_0} p_{i_0 i_1} \cdots p_{i_{n-1} i_n}$ for all $n \in \mathbf{N}$.
4. Define X_n by saying that $X_n(\omega) = i_n$ if $\omega \in I_{i_0 i_1 \ldots i_{n-1} i_n}^{(n)}$ for some choice of $i_0, \ldots, i_{n-1} \in S$.

Then it is easily verified that the random variables $\{X_n\}$ have the desired properties.   ∎

We thus see that, as in Theorem 7.1.1, an "old friend" (in this case Lebesgue measure on $[0,1]$) was able to serve as a probability space on which to define important random variables.

8.2. Transience, recurrence, and irreducibility.

In this subsection, we consider some fundamental notions related to Markov chains. For simplicity, we shall assume that $\nu_i > 0$ for all $i \in S$, and shall write $\mathbf{P}_i(\cdots)$ as shorthand for the conditional probability $\mathbf{P}(\cdots | X_0 = i)$. Intuitively, $\mathbf{P}_i(A)$ stands for the probability that the event A would have occurred, had the Markov chain been started at the state i. We shall also write $\mathbf{E}_i(\cdots)$ for expected values with respect to $\mathbf{P}_i$.

To proceed, we define, for $i, j \in S$ and $n \in \mathbf{N}$, the probabilities

$$f_{ij}^{(n)} = \mathbf{P}_i(X_n = j, \text{ but } X_m \neq j \text{ for } 1 \leq m \leq n-1);$$

$$f_{ij} = \mathbf{P}_i(\exists n \geq 1; X_n = j) = \sum_{n=1}^{\infty} f_{ij}^{(n)}.$$

That is, $f_{ij}^{(n)}$ is the probability, starting from i, that we *first* hit j at the time n; f_{ij} is the probability, starting from i, that we *ever* hit j.

A state $i \in S$ is called *recurrent* (or *persistent*) if $f_{ii} = 1$, i.e. if starting from i we will certainly eventually return to i. It is called *transient* if it is not recurrent, i.e. if $f_{ii} < 1$. Recurrence and transience are very important concepts in Markov chain theory, and we prove some results about them here. (Recall that *i.o.* stands for *infinitely often*.)

Theorem 8.2.1.   Let $\{X_n\}$ be a Markov chain, on a state space S. Let $i \in S$. Then i is transient if and only if $\mathbf{P}_i(X_n = i \text{ i.o.}) = 0$ if and only if $\sum_{n=1}^{\infty} p_{ii}^{(n)} < \infty$. On the other hand, i is recurrent if and only if $\mathbf{P}_i(X_n = i \text{ i.o.}) = 1$ if and only if $\sum_{n=1}^{\infty} p_{ii}^{(n)} = \infty$.

To prove this theorem, we begin with a lemma.

Lemma 8.2.2.   We have $\mathbf{P}_i(\#\{n \geq 1; X_n = j\} \geq k) = f_{ij}(f_{jj})^{k-1}$, for $k = 1, 2, \ldots$. In words, the probability that, starting from i, we eventually hit the state j at least k times, is given by $f_{ij}(f_{jj})^{k-1}$.

Proof.   Starting from i, to hit j at least k times is equivalent to first hitting j once (starting from i), then to return to j at least $k - 1$ more times (each time starting from j). The result follows. ∎

Proof of Theorem 8.2.1. By continuity of probabilities, and by Lemma 8.2.2, we have that

$$\mathbf{P}_i(X_n = i \text{ i.o.}) = \lim_{k \to \infty} \mathbf{P}_i \left(\#\{n \geq 1 \, ; \, X_n = i\} \geq k \right)$$

$$= \lim_{k \to \infty} (f_{ii})^k = \begin{cases} 0, & f_{ii} < 1 \\ 1, & f_{ii} = 1. \end{cases}$$

This proves the first equivalence to each of transience and recurrence.

Also, using first countable linearity, and then Proposition 4.2.9, and then Lemma 8.2.2, we compute that

$$\begin{aligned}
\sum_{n=1}^{\infty} p_{ii}^{(n)} &= \sum_{n=1}^{\infty} \mathbf{P}_i(X_n = i) \\
&= \sum_{n=1}^{\infty} \mathbf{E}_i(\mathbf{1}_{X_n = i}) \\
&= \mathbf{E}_i \left(\sum_{n=1}^{\infty} \mathbf{1}_{X_n = i} \right) \\
&= \mathbf{E}_i \left(\#\{n \geq 1; X_n = i\} \right) \\
&= \sum_{k=1}^{\infty} \mathbf{P}_i \left(\#\{n \geq 1; X_n = i\} \geq k \right) \\
&= \sum_{k=1}^{\infty} (f_{ii})^k \\
&= \begin{cases} \frac{f_{ii}}{1 - f_{ii}} < \infty, & f_{ii} < 1 \\ \infty, & f_{ii} = 1, \end{cases}
\end{aligned}$$

thus proving the remaining two equivalences. ∎

As an application of this theorem, we consider simple symmetric random walk (cf. Subsection 7.2 or Example 8.0.1, with $X_0 = 0$ and $p = \frac{1}{2}$). We recall that here $S = \mathbf{Z}$, and for any $i \in \mathbf{Z}$ we have $p_{ii}^{(n)} = \binom{n}{n/2} \left(\frac{1}{2}\right)^n = \frac{n!}{((n/2)!)^2 2^n}$ for n even (with $p_{ii}^{(n)} = 0$ for n odd). Using Sterling's approximation, which states that $n! \sim \left(\frac{n}{e}\right)^n \sqrt{2\pi n}$ as $n \to \infty$, we compute that for large even n, we have $p_{ii}^{(n)} \sim \sqrt{2/\pi n}$. Hence, we see (cf. (A.3.7)) that $\sum_{n=1}^{\infty} p_{ii}^{(n)} = \infty$. Therefore, by Theorem 8.2.1, simple symmetric random walk is recurrent.

On the other hand, if $p \neq \frac{1}{2}$ for simple random walk, then $p_{ii}^{(n)} = \binom{n}{n/2} (p(1-p))^{n/2}$ with $p(1-p) < \frac{1}{4}$. Sterling then gives for large even n that $p_{ii}^{(n)} \sim [4p(1-p)]^{n/2} \sqrt{2/\pi n}$, with $4p(1-p) < 1$. It follows that $\sum_{n=1}^{\infty} p_{ii}^{(n)} < \infty$, so that simple *asymmetric* random walk is *not* recurrent.

In higher dimensions, suppose that $S = \mathbf{Z}^d$, with each of the d coordinates independently following simple symmetric random walk. Then clearly $p_{ii}^{(n)}$ for this process is simply the dth power of the corresponding quantity for ordinary (one-dimensional) simple symmetric random walk. Hence, for large even n, we have $p_{ii}^{(n)} \sim (2/\pi n)^{d/2}$ in this case. It follows (cf. (A.3.7)) that $\sum_{n=1}^{\infty} p_{ii}^{(n)} = \infty$ if and only if $d \leq 2$. That is, higher-dimensional simple symmetric random walk is recurrent in dimensions 1 and 2, but transient in dimensions ≥ 3, a somewhat counter-intuitive result.

Finally, we note that for irreducible Markov chains, somewhat more can be said. A Markov chain is *irreducible* if $f_{ij} > 0$ for all $i, j \in S$, i.e. if it is possible for the chain to move from any state to any other state. Equivalently, the Markov chain is irreducible if for any $i, j \in S$, there is $r \in \mathbf{N}$ with $p_{ij}^{(r)} > 0$. (A chain is *reducible* if it is not irreducible, i.e. if $f_{ij} = 0$ for some $i, j \in S$.) We can now prove

Theorem 8.2.3. *Let $\{p_{ij}\}_{i,j \in S}$ be the transition probabilities for an irreducible Markov chain on a state space S. Then the following are equivalent:*

(1) *There is $k \in S$ with $f_{kk} = 1$, i.e. with k recurrent.*

(2) *For all $i, j \in S$, we have $f_{ij} = 1$ (so, in particular, all states are recurrent).*

(3) *There are $k, \ell \in S$ with $\sum_{n=1}^{\infty} p_{k\ell}^{(n)} = \infty$.*

(4) *For all $i, j \in S$, we have $\sum_{n=1}^{\infty} p_{ij}^{(n)} = \infty$.*

If any of (1)–(4) hold, we say that the Markov chain itself is *recurrent*.

Proof. That (2) implies (1) is immediate.

That (1) implies (3) follows immediately from Theorem 8.2.1.

To show that (3) implies (4): Assume that (3) holds, and let $i, j \in S$. By irreducibility, there are $m, r \in \mathbf{N}$ with $p_{ik}^{(m)} > 0$ and $p_{\ell j}^{(r)} > 0$. But then, since $p_{ij}^{(m+n+r)} \geq p_{ik}^{(m)} p_{k\ell}^{(n)} p_{\ell j}^{(r)}$, we have $\sum_n p_{ij}^{(n)} \geq p_{ik}^{(m)} p_{\ell j}^{(r)} \sum_n p_{k\ell}^{(n)} = \infty$, as claimed.

To show that (4) implies (2): Suppose, to the contrary of (2), that $f_{ij} < 1$ for some $i, j \in S$. Then $1 - f_{jj} \geq \mathbf{P}_j(\tau_i < \tau_j)(1 - f_{ij}) > 0$. (Here $\mathbf{P}_j(\tau_i < \tau_j)$ stands for the probability, starting from j, that we hit i before returning to j; and it is positive by irreducibility.) Hence, $f_{jj} < 1$, so by Theorem 8.2.1 we have $\sum_n p_{jj}^{(n)} < \infty$, contradicting (4). ∎

For example, for simple symmetric random walk, since $\sum_n p_{ii}^{(n)} = \infty$, this theorem says that from *any* state i, the walk will (with probability 1) eventually reach *any* other state j. Thus, the walk will keep on wondering from state to state forever; this is related to "fluctuation theory" for random walks. In particular, this implies that for simple symmetric random walk, with probability 1 we have $\limsup_n X_n = \infty$ and $\liminf_n X_n = -\infty$.

Such considerations are related to the remarkable *law of the iterated logarithm*. Let $X_n = Z_1 + \ldots + Z_n$ define a random walk, where $Z_1, Z_2, \ldots$

are i.i.d. with mean 0 and variance 1. (For example, perhaps $\{X_n\}$ is simple symmetric random walk.) Then it is a fact that with probability 1, $\limsup_n (X_n / \sqrt{2\, n \log\log n}) = 1$. In other words, for any $\epsilon > 0$, we have $\mathbf{P}(X_n \geq (1+\epsilon)\sqrt{2\, n \log\log n} \; i.o.) = 0$ and $\mathbf{P}(X_n \geq (1-\epsilon)\sqrt{2\, n \log\log n} \; i.o.) = 1$. This gives extremely precise information about how the peaks of $\{X_n\}$ grow for large n. For a proof see e.g. Billingsley (1995, Theorem 9.5).

8.3. Stationary distributions and convergence.

Given a Markov chain on a state space S, with transition probabilities $\{p_{ij}\}$, let $\{\pi_i\}_{i \in S}$ be a distribution on S (i.e., $\pi_i \geq 0$ for all $i \in S$, and $\sum_{i \in S} \pi_i = 1$). Then $\{\pi_i\}_{i \in S}$ is *stationary* for the Markov chain if $\sum_{i \in S} \pi_i p_{ij} = \pi_j$ for all $j \in S$. (In matrix form, $[\pi][\mathbf{p}] = [\pi]$.)

What this means is that if we start the Markov chain in the distribution $\{\pi_i\}$ (i.e., $\mathbf{P}[X_0 = i] = \pi_i$ for all $i \in S$), then one time unit later the distribution will *still* be $\{\pi_i\}$ (i.e., $\mathbf{P}[X_1 = i] = \pi_i$ for all $i \in S$); this is the reason for the terminology "stationary". It follows immediately by induction that the chain will still be in the same distribution $\{\pi_i\}$ any number n steps later (i.e., $\mathbf{P}[X_n = i] = \pi_i$ for all $i \in S$). Equivalently, $\sum_{i \in S} \pi_i p_{ij}^{(n)} = \pi_j$ for any $n \in \mathbf{N}$. (In matrix form this is even clearer: $[\pi][\mathbf{p}]^n = [\pi]$.)

Exercise 8.3.1. A Markov chain is said to be *reversible* with respect to a distribution $\{\pi_i\}$ if, for all $i, j \in S$, we have $\pi_i p_{ij} = \pi_j p_{ji}$. Prove that, if a Markov chain is reversible with respect to $\{\pi_i\}$, then $\{\pi_i\}$ is a stationary distribution for the chain.

For example, for random walk on $\mathbf{Z}/(d)$ as in Example 8.0.3, it is computed that the uniform distribution, given by $\pi_i = 1/d$ for $i = 0, 1, \ldots, d-1$, is a stationary distribution. For Ehrenfest's Urn (Example 8.0.4), it is computed that the binomial distribution, given by $\pi_i = \binom{d}{i}\frac{1}{2^d}$ for $i = 0, 1, \ldots, d$, is a stationary distribution.

Exercise 8.3.2. Verify these last two statements.

Now, given a Markov chain with a stationary distribution, one might expect that if the chain is run for a long time (i.e. $n \to \infty$), that the probability of being at a particular state $j \in S$ might converge to π_j, regardless of the initial state chosen. That is, one might expect that $\lim_{n \to \infty} \mathbf{P}_i(X_n = j) = \pi_j$ for any $i, j \in S$. This is not true in complete generality, as the following two examples show. However, we shall see in Theorem 8.3.10 below that this is indeed true for many Markov chains.

For a first example, suppose $S = \{1, 2\}$, and that the transition probabilities are given by

$$(p_{ij}) = \begin{pmatrix} 1 & 0 \\ 0 & 1 \end{pmatrix} .$$

That is, this Markov chain never moves at all! Hence, *any* distribution is stationary for this chain. In particular, we could take $\pi_1 = \pi_2 = \frac{1}{2}$ as a stationary distribution. On the other hand, we clearly have $\mathbf{P}_1(X_n = 1) = 1$ for all $n \in \mathbf{N}$, which certainly does not approach $\frac{1}{2}$. Thus, this Markov chain does not converge to the stationary distribution $\{\pi_i\}$. In fact, this Markov chain is clearly reducible (i.e., not irreducible), which is the obstacle to convergence.

For a second example, suppose again that $S = \{1, 2\}$, and that the transition probabilities are given this time by

$$(p_{ij}) = \begin{pmatrix} 0 & 1 \\ 1 & 0 \end{pmatrix} .$$

Again we may take $\pi_1 = \pi_2 = \frac{1}{2}$ as a stationary distribution (in fact, this time the stationary distribution is unique). Furthermore, this Markov chain is irreducible. On the other hand, we have $\mathbf{P}_1(X_n = 1) = 1$ for n even, and $\mathbf{P}_1(X_n = 1) = 0$ for n odd. Hence, again we do *not* have $\mathbf{P}_1(X_n = 1) \to \frac{1}{2}$. (On the other hand, the Cesàro averages of $\mathbf{P}_1(X_n = 1)$, i.e. $\frac{1}{n} \sum_{i=1}^{n} \mathbf{P}_1(X_i = 1)$, do indeed converge to $\frac{1}{2}$, which is not a coincidence.) Here the obstacle to convergence is that the Markov chain is "periodic", with period 2, as we now discuss.

Definition 8.3.3. Given Markov chain transitions $\{p_{ij}\}$ on a state space S, and a state $i \in S$, the *period* of i is the greatest common divisor of the set $\{n \geq 1; p_{ii}^{(n)} > 0\}$.

That is, the period of i is the g.c.d. of the times at which it is possible to travel from i to i. For example, if the period of i is 2, then this means it is only possible to travel from i to i in an even number of steps. (Such was the case for the second example above.) Clearly, if the period of a state is greater than 1, then this will be an obstacle to convergence to a stationary distribution. This prompts the following definition.

Definition 8.3.4. A Markov chain is *aperiodic* if the period of each state is 1. (A chain which is not aperiodic is said to be *periodic*.)

Before proceeding, we note a fact about periods that makes aperiodicity easier to verify.

Lemma 8.3.5. *Let i and j be two states of a Markov chain, and suppose that $f_{ij} > 0$ and $f_{ji} > 0$ (i.e., i and j "communicate"). Then the periods of i and of j are equal.*

Proof. Since $f_{ij} > 0$ and $f_{ji} > 0$, there are $r, s \in \mathbf{N}$ with $p_{ij}^{(r)} > 0$ and $p_{ji}^{(s)} > 0$. Since $p_{ii}^{(r+n+s)} \geq p_{ij}^{(r)} p_{jj}^{(n)} p_{ji}^{(s)}$, this implies that

$$p_{ii}^{(r+n+s)} > 0 \qquad \text{whenever} \qquad p_{jj}^{(n)} > 0. \qquad (8.3.6)$$

Now, if we let the periods of i and j be t_i and t_j, respectively, then (8.3.6) with $n = 0$ implies that t_i divides $r + s$. Then, for any $n \in \mathbf{N}$ with $p_{jj}^{(n)} > 0$, (8.3.6) implies that t_i divides $r + n + s$, hence that t_i divides n. That is, t_i is a common divisor of $\{n \in \mathbf{N};\ p_{jj}^{(n)} > 0\}$. Since t_j is the *greatest* common divisor of this set, we must have $t_i \leq t_j$. Similarly, $t_j \leq t_i$, so we must have $t_i = t_j$. ∎

This immediately implies

Corollary 8.3.7. *If a Markov chain is irreducible, then all of its states have the same period.*

Hence, for an irreducible Markov chain, it suffices to check aperiodicity at any single state.

We shall prove in Theorem 8.3.10 below that all Markov chains which are irreducible and aperiodic, and have a stationary distribution, do in fact converge to it. Before doing so, we require two further lemmas, which give more concrete implications of irreducibility and aperiodicity.

Lemma 8.3.8. *If a Markov chain is irreducible, and has a stationary distribution $\{\pi_i\}$, then it is recurrent.*

Proof. Suppose to the contrary that the chain is not recurrent. Then, by Theorem 8.2.3, we have $\sum_n p_{ij}^{(n)} < \infty$ for all states i and j; in particular, $\lim_{n \to \infty} p_{ij}^{(n)} = 0$. Now, since $\{\pi_i\}$ is stationary, we have $\pi_j = \sum_i \pi_i p_{ij}^{(n)}$ for all $n \in \mathbf{N}$. Hence, letting $n \to \infty$, we see (formally, by using the M-test, cf. Proposition A.4.8) that we must have $\pi_j = 0$, for all states j. This contradicts the fact that $\sum_j \pi_j = 1$. ∎

Remark. The *converse* to Lemma 8.3.8 is false, in the sense that just because an irreducible Markov chain is recurrent, it might not have a stationary distribution (e.g. this is the case for simple symmetric random walk).

Lemma 8.3.9. *If a Markov chain is irreducible and aperiodic, then for each pair (i, j) of states, there is a number $n_0 = n_0(i, j)$, such that $p_{ij}^{(n)} > 0$ for all $n \geq n_0$.*

Proof. Fix states i and j. Let $T = \{n \geq 1; p_{ii}^{(n)} > 0\}$. Then, by aperiodicity, $gcd(T) = 1$. Hence, we can find $m \in \mathbf{N}$, and $k_1, \ldots, k_m \in T$, and integers $b_1, \ldots, b_m$, such that $k_1 b_1 + \ldots + k_m b_m = 1$. We furthermore choose any $a \in T$, and also (by irreducibility) choose $c \in \mathbf{N}$ with $p_{ij}^{(c)} > 0$. These values shall be the key to defining n_0.

We now set $M = k_1 |b_1| + \ldots + k_m |b_m|$ (i.e., a sum that without the absolute value signs would be 1), and define $n_0 = n_0(i, j) = aM + c$. Then, if $n \geq n_0$, then letting $r = \lfloor (n - c)/a \rfloor$, we can write $n = c + ra + s$ where $0 \leq s < a$ and $r \geq M$. We then observe that, since $\sum_{\ell=1}^{m} b_\ell k_\ell = 1$ and $\sum_{\ell=1}^{m} |b_\ell| k_\ell = M$, we have

$$n = (r - M)a + \sum_{\ell=1}^{m} (a|b_\ell| + sb_\ell) k_\ell + c,$$

where the quantities in brackets are non-negative. Hence, recalling that $a, k_\ell \in T$, and that $p_{ij}^{(c)} > 0$, we have that

$$p_{ij}^{(n)} \geq \left(p_{ii}^{(a)}\right)^{r-M} \left[\prod_{\ell=1}^{m} \left(p_{ii}^{(k_\ell)}\right)^{a|b_\ell|+sb_\ell}\right] p_{ij}^{(c)} > 0,$$

as required.   ∎

We are now able to prove our main Markov chain convergence theorem.

Theorem 8.3.10. *If a Markov chain is irreducible and aperiodic, and has a stationary distribution $\{\pi_i\}$, then for all states i and j, we have $\lim_{n \to \infty} \mathbf{P}_i(X_n = j) = \pi_j$.*

Proof. The proof uses the method of "coupling". Let the original Markov chain have state space S, and transition probabilities $\{p_{ij}\}$. We define a *new* Markov chain $\{(X_n, Y_n)\}_{n=0}^{\infty}$, having state space $\overline{S} = S \times S$, and transition probabilities given by $\overline{p}_{(ij),(k\ell)} = p_{ik} p_{j\ell}$. That is, our new Markov chain has two coordinates, each of which is an independent copy of the original Markov chain.

It follows immediately that the distribution on $\overline{S}$ given by $\overline{\pi}_{(ij)} = \pi_i \pi_j$ (i.e., a product of two probabilities in the original stationary distribution) is in fact a stationary distribution for our new Markov chain. Furthermore, from Lemma 8.3.9 above, we see that our new Markov chain will again be irreducible and aperiodic (indeed, we have $\overline{p}^{(n)}_{(ij),(k\ell)} > 0$ whenever $n \geq \max(n_0(i,k), n_0(j,\ell))$). Hence, from Lemma 8.3.8, we see that our new Markov chain is in fact recurrent. This is the key to what follows.

To complete the proof, we choose $i_0 \in S$, and set $\tau = \inf\{n \geq 1; X_n = Y_n = i_0\}$. Note that, for $m \leq n$, the quantities $\mathbf{P}_{(ij)}(\tau = m, X_n = k)$ and $\mathbf{P}_{(ij)}(\tau = m, Y_n = k)$ are equal; indeed, they both equal $\mathbf{P}_{(ij)}(\tau = m)p^{(n-m)}_{i_0,k}$. Hence, for any states i, j, and k, we have that

$$
\begin{aligned}
\left| p^{(n)}_{ik} - p^{(n)}_{jk} \right| &= \left| \mathbf{P}_{(ij)}(X_n = k) - \mathbf{P}_{(ij)}(Y_n = k) \right| \\
&= \left| \sum_{m=1}^{n} \mathbf{P}_{(ij)}(X_n = k, \ \tau = m) + \mathbf{P}_{(ij)}(X_n = k, \ \tau > n) \right. \\
&\quad \left. - \sum_{m=1}^{n} \mathbf{P}_{(ij)}(Y_n = k, \ \tau = m) - \mathbf{P}_{(ij)}(Y_n = k, \ \tau > n) \right| \\
&= \left| \mathbf{P}_{(ij)}(X_n = k, \ \tau > n) - \mathbf{P}_{(ij)}(Y_n = k, \ \tau > n) \right| \\
&\leq \mathbf{P}_{(ij)}(\tau > n)
\end{aligned}
$$

(where the inequality follows since we are considering a difference of two positive quantities, each of which is $\leq \mathbf{P}_{(ij)}(\tau > n)$). Now, since the new Markov chain is irreducible and recurrent, it follows from Theorem 8.2.3 that $\overline{f}_{(ij),(i_0 i_0)} = 1$. That is, with probability 1, the chain will eventually hit the state (i_0, i_0), in which case $\tau < \infty$. Hence, as $n \to \infty$, we have $\mathbf{P}_{(ij)}(\tau > n) \to 0$, so that $\left| p^{(n)}_{ik} - p^{(n)}_{jk} \right| \to 0$.

On the other hand, by stationarity, we have for the original Markov chain that

$$
\left| p^{(n)}_{ij} - \pi_j \right| = \left| \sum_{k \in S} \pi_k \left(p^{(n)}_{ij} - p^{(n)}_{kj} \right) \right| \leq \sum_{k \in S} \pi_k \left| p^{(n)}_{ij} - p^{(n)}_{kj} \right|,
$$

and we see (again, by the M-test) that this converges to 0 as $n \to \infty$. Since $p^{(n)}_{ij} = \mathbf{P}_i(X_n = j)$, this proves the theorem. ∎

Finally, we note that Theorem 8.3.10 remains true if the chain begins in some other initial distribution $\{\nu_i\}$ besides those with $\nu_i = 1$ for some i:

Corollary 8.3.11. *If a Markov chain is irreducible and aperiodic, and has a stationary distribution $\{\pi_i\}$, then regardless of the initial distribution, for all $j \in S$, we have $\lim_{n\to\infty} \mathbf{P}(X_n = j) = \pi_j$.*

Proof.   Using the M-test and Theorem 8.3.10, we have

$$
\begin{aligned}
\lim_{n\to\infty} \mathbf{P}(X_n = j) &= \lim_{n\to\infty} \sum_{i \in S} \mathbf{P}(X_0 = i,\ X_n = j) \\
&= \lim_{n\to\infty} \sum_{i \in S} \mathbf{P}(X_0 = i)\,\mathbf{P}(X_n = j \mid X_0 = i) \\
&= \lim_{n\to\infty} \sum_{i \in S} \nu_i\,\mathbf{P}_i(X_n = j) \\
&= \sum_{i \in S} \nu_i \lim_{n\to\infty} \mathbf{P}_i(X_n = j) \\
&= \sum_{i \in S} \nu_i\,\pi_j \\
&= \pi_j,
\end{aligned}
$$

which gives the result.   ∎

8.4. Existence of stationary distributions.

Theorem 8.3.10 gives powerful information about the convergence of irreducible and aperiodic Markov chains. However, this theorem requires the existence of a stationary distribution $\{\pi_i\}$. It is reasonable to ask for conditions under which a stationary distribution will exist. We consider that question here.

Given a Markov chain $\{X_n\}$ on a state space S, and given a state $i \in S$, we define the *mean return time* m_i by $m_i = \mathbf{E}_i\,(\inf\{n \geq 1; X_n = i\})$. That is, m_i is the expected time to return to the state i. We always have $m_i \geq 1$. If i is transient, then with positive probability we will have $\inf\{n \geq 1; X_n = i\} = \infty$, so of course we will have $m_i = \infty$. On the other hand, if i is recurrent, then we shall call i *null recurrent* if $m_i = \infty$, and shall call i *positive recurrent* if $m_i < \infty$. (The names come from considering $1/m_i$ rather than m_i.)

The main theorem of this subsection is

Theorem 8.4.1.   *If a Markov chain is irreducible, and if each state i of the Markov chain is positive recurrent with (finite) mean return time m_i, then the Markov chain has a unique stationary distribution $\{\pi_i\}$, given by $\pi_i = 1/m_i$ for each state i.*

This theorem is rather surprising. It is not immediately clear that the mean return times m_i have *anything* to do with a stationary distribution; it is even less expected that they provide a precise formula for the stationary distribution values. The proof of the theorem relies heavily on the following lemma.

Lemma 8.4.2.   *Let $G_n(i,j) = \mathbf{E}_i\,(\#\{\ell;\ 1 \leq \ell \leq n,\ X_\ell = j\}) = \sum_{\ell=1}^{n} p_{ij}^{(\ell)}$. Then for an irreducible recurrent Markov chain,*

$$
\lim_{n\to\infty} \frac{G_n(i,j)}{n} = \frac{1}{m_j}
$$

for any states i and j.

Proof. Let T_j^r be the time of the rth hit of the state j. Then

$$T_j^r = T_j^1 + (T_j^2 - T_j^1) + \ldots + (T_j^r - T_j^{r-1});$$

here the $r - 1$ terms in brackets are i.i.d. and have mean m_j. Hence, from the Strong Law of Large Numbers (second version, Theorem 5.4.4), we have that $\lim_{r \to \infty} \frac{T_j^r}{r} = m_j$ with probability 1.

Now, for $n \in \mathbf{N}$, let $r(n) = \#\{\ell; \ 1 \leq \ell \leq n, \ X_\ell = j\}$. Then $\lim_{n \to \infty} r(n) = \infty$ with probability 1 by recurrence. Also clearly $T_j^{r(n)} \leq n < T_j^{r(n)+1}$, so that

$$\frac{T_j^{r(n)}}{r(n)} \leq \frac{n}{r(n)} < \frac{T_j^{r(n)+1}}{r(n)}.$$

Hence, we must have $\lim_{n \to \infty} \frac{n}{r(n)} = m_j$ with probability 1 as well. Therefore, with probability 1 we have $\lim_{n \to \infty} \frac{r(n)}{n} = 1/m_j$.

On the other hand, $G_n(i, j) = \mathbf{E}_i(r(n))$, and furthermore $0 \leq \frac{r(n)}{n} \leq 1$. Hence, by the bounded convergence theorem,

$$\lim_{n \to \infty} \frac{G_n(i, j)}{n} = \lim_{n \to \infty} \mathbf{E}_i\left(\frac{r(n)}{n}\right) = 1/m_j,$$

as claimed. ∎

Remark 8.4.3. We note that this proof goes through without change in the case $m_j = \infty$, so that $\lim_{n \to \infty} \frac{G_n(i,j)}{n} = 0$ in that case.

Proof of Theorem 8.4.1. We begin by proving the uniqueness. Indeed, suppose that $\sum_i \alpha_i p_{ij} = \alpha_j$ for all states j, for some probability distribution $\{\alpha_i\}$. Then, by induction, $\sum_i \alpha_i p_{ij}^{(t)} = \alpha_j$ for all $t \in \mathbf{N}$, so that also $\frac{1}{n} \sum_{t=1}^n \sum_i \alpha_i p_{ij}^{(t)} = \alpha_j$. Hence,

$$
\begin{aligned}
\alpha_j &= \lim_{n \to \infty} \tfrac{1}{n} \sum_{t=1}^n \sum_i \alpha_i p_{ij}^{(t)} \\
&= \sum_i \alpha_i \lim_{n \to \infty} \tfrac{1}{n} \sum_{t=1}^n p_{ij}^{(t)} \qquad \text{by the M-test} \\
&= \sum_i \alpha_i \tfrac{1}{m_j} \qquad \text{by the Lemma} \\
&= 1/m_j.
\end{aligned}
$$

That is, $\alpha_j = 1/m_j$ is the only possible stationary distribution.

Now, fix any state z. Then for all $t \in \mathbf{N}$, we have $\sum_j p_{zj}^{(t)} = 1$. Hence, if we set $C = \sum_j 1/m_j$, then using the Lemma and (A.4.9), we have

$$C = \sum_j \lim_{n \to \infty} \frac{1}{n} \sum_{t=1}^{n} p_{zj}^{(t)} \leq \lim_{n \to \infty} \sum_j \frac{1}{n} \sum_{t=1}^{n} p_{zj}^{(t)} = 1.$$

In particular, $C < \infty$.

Again fix any state z. Then for all states j, and for all $t \in \mathbf{N}$, we have $\sum_i p_{zi}^{(t)} p_{ij} = p_{zj}^{(t+1)}$. Hence, again by the Lemma and (A.4.9), we have

$$1/m_j = \lim_{n \to \infty} \frac{1}{n} \sum_{t=1}^{n} p_{zj}^{(t+1)} = \lim_{n \to \infty} \sum_i \frac{1}{n} \sum_{t=1}^{n} p_{zi}^{(t)} p_{ij} \geq \sum_i \frac{1}{m_i} p_{ij}. \quad (8.4.4)$$

But then summing both sides over all states j, and recalling that $\sum_j \frac{1}{m_j} = C < \infty$, we see that we must have equality in (8.4.4), i.e. we must have $\frac{1}{m_j} = \sum_i \frac{1}{m_i} p_{ij}$. But this means that the probability distribution defined by $\pi_i = 1/Cm_i$ must be a stationary distribution for the Markov chain! Finally, by uniqueness, we must have $\pi_j = 1/m_j$ for each state j, i.e. we must have $C = 1$. This completes the proof of the theorem. ∎

On the other hand, states which are *not* positive recurrent cannot contribute to a stationary distribution:

Proposition 8.4.5. *If a Markov chain has a stationary distribution $\{\pi_i\}$, and if a state j is not positive recurrent (i.e., satisfies $m_j = \infty$), then we must have $\pi_j = 0$.*

Proof. We have that $\sum_i \pi_i p_{ij}^{(t)} = \pi_j$ for all $t \in \mathbf{N}$. Hence, using the M-test and also Lemma 8.4.2 and Remark 8.4.3, we have

$$\pi_j = \lim_{n \to \infty} \frac{1}{n} \sum_i \pi_i \sum_{t=1}^{n} p_{ij}^{(t)} = \sum_i \pi_i \lim_{n \to \infty} \frac{1}{n} \sum_{t=1}^{n} p_{ij}^{(t)} = \sum_i (\pi_i)(0) = 0,$$

as claimed. ∎

Corollary 8.4.6. *If a Markov chain has no positive recurrent states, then it does not have a stationary distribution.*

Proof. Suppose to the contrary that it did have a stationary distribution $\{\pi_i\}$. Then from the above proposition, we would necessarily have $\pi_j = 0$

for each state j, contradicting the fact that $\sum_j \pi_j = 1$. ∎

Theorem 8.4.1 and Corollary 8.4.6 provide clear information about Markov chains where all states are positive recurrent, or where none are, respectively. One could still wonder about chains which have some positive recurrent and some non-positive-recurrent states. We now show that, for irreducible Markov chains, this cannot happen. The statement is somewhat analogous to that of Lemma 8.3.5.

Proposition 8.4.7. *Let i and j be two states of a Markov chain. Suppose that $f_{ij} > 0$ and $f_{ji} > 0$ (i.e., the states i and j communicate). Then if i is positive recurrent, then j is also positive recurrent.*

Proof. Find $r, t \in \mathbf{N}$ with $p_{ji}^{(r)} > 0$ and $p_{ij}^{(t)} > 0$. Then by Lemma 8.4.2 and Remark 8.4.3,

$$\frac{1}{m_j} = \lim_{n \to \infty} \frac{1}{n} \sum_{m=1}^{n} p_{jj}^{(m)} \geq \lim_{n \to \infty} \frac{1}{n} \sum_{m=r+t}^{n} p_{ji}^{(r)} p_{ii}^{(m-r-t)} p_{ij}^{(t)} = \frac{p_{ji}^{(r)} p_{ij}^{(t)}}{m_i} > 0,$$

so that $m_j < \infty$. ∎

Corollary 8.4.8. *For an irreducible Markov chain, either all states are positive recurrent or none are.*

Combining this corollary with Theorem 8.4.1 and Corollary 8.4.6, we see that

Theorem 8.4.9. *Consider an irreducible Markov chain. Then either (a) all states are positive recurrent, and there exists a unique stationary distribution, given by $\pi_j = 1/m_j$, to which (assuming aperiodicity) $\mathbf{P}_i(X_n = j)$ converges as $n \to \infty$; or (b) no states are positive recurrent, and there does not exist a stationary distribution.*

For example, consider simple symmetric random walk, with state space the integers $\mathbf{Z}$. It is clear that m_j must be the same for all states j (i.e. no state is any "different" from any other state). Hence, it is impossible that $\sum_{j \in \mathbf{Z}} 1/m_j = 1$. Thus, simple symmetric random walk cannot fall into category (a) above. Hence, simple symmetric random walk falls into category (b), and does not have a stationary distribution; in fact, it is null recurrent.

Finally, we observe that all irreducible Markov chains on *finite* state spaces necessarily fall into category (a) above:

Proposition 8.4.10. *For an irreducible Markov chain on a finite state space, all states are positive recurrent (and hence a unique stationary distribution exists).*

Proof. Fix a state i. Write $h_{ji}^{(m)} = \mathbf{P}_j(X_t = i$ for some $1 \leq t \leq m) = \sum_{n=1}^m f_{ji}^{(n)}$. Then $\lim_{m \to \infty} h_{ji}^{(m)} = f_{ji} > 0$, for each state j. Hence, since the state space is finite, we can find $m \in \mathbf{N}$ and $\delta > 0$ such that $h_{ji}^{(m)} \geq \delta$ for all states j.

But then we must have $1 - h_{ii}^{(n)} \leq (1 - \delta)^{\lfloor n/m \rfloor}$, so that letting $\tau_i = \inf\{n \geq 1; X_n = i\}$, we have by Proposition 4.2.9 that

$$m_i = \sum_{n=0}^{\infty} \mathbf{P}_i(\tau_i \geq n+1) = \sum_{n=0}^{\infty}(1 - h_{ii}^{(n)}) \leq \sum_{n=0}^{\infty}(1 - \delta)^{\lfloor n/m \rfloor} = \frac{m}{\delta} < \infty. \ \blacksquare$$

8.5. Exercises.

Exercise 8.5.1. Consider a discrete-time, time-homogeneous Markov chain with state space $S = \{1, 2\}$, and transition probabilities given by

$$p_{11} = a, \quad p_{12} = 1 - a, \quad p_{21} = 1, \quad p_{22} = 0.$$

For each $0 \leq a \leq 1$,
(a) Compute $p_{ij}^{(n)} = P(X_n = j \mid X_0 = i)$ for each $i, j \in \mathcal{X}$ and $n \in \mathbf{N}$.
(b) Classify each state as recurrent or transient.
(c) Find all stationary distributions for this Markov chain.

Exercise 8.5.2. For any $\epsilon > 0$, give an example of an irreducible Markov chain on a countably infinite state space, such that $|p_{ij} - p_{ik}| \leq \epsilon$ for all states i, j, and k.

Exercise 8.5.3. For an arbitrary Markov chain on a state space consisting of exactly d states, find (with proof) the largest possible positive integer N such that for some states i and j, we have $p_{ij}^{(N)} > 0$ but $p_{ij}^{(n)} = 0$ for all $n < N$.

Exercise 8.5.4. Given Markov chain transition probabilities $\{p_{ij}\}_{i,j \in S}$ on a state space S, call a subset $C \subseteq S$ *closed* if $\sum_{j \in C} p_{ij} = 1$ for each $i \in C$. Prove that a Markov chain is irreducible if and only if it has no closed subsets (aside from the empty set and S itself).

Exercise 8.5.5. Suppose we modify Example 8.0.3 so the chain moves one unit clockwise with probability r, or one unit counter-clockwise with probability $1 - r$, for some $0 < r < 1$. That is, $S = \{0, 1, 2, \ldots, d - 1\}$ and

$$p_{ij} = \begin{cases} r, & j = i + 1 (\mathrm{mod}\ d) \\ 1 - r, & j = i - 1 (\mathrm{mod}\ d) \\ 0, & \text{otherwise} . \end{cases}$$

Find (with explanation) all stationary distributions of this Markov chain.

Exercise 8.5.6. Consider the Markov chain with state space $S = \{1, 2, 3\}$ and transition probabilities $p_{12} = p_{23} = p_{31} = 1$. Let $\pi_1 = \pi_2 = \pi_3 = 1/3$.
(a) Determine whether or not the chain is irreducible.
(b) Determine whether or not the chain is aperiodic.
(c) Determine whether or not the chain is reversible with respect to $\{\pi_i\}$.
(d) Determine whether or not $\{\pi_i\}$ is a stationary distribution.
(e) Determine whether or not $\lim_{n \to \infty} p_{11}^{(n)} = \pi_1$.

Exercise 8.5.7. Give an example of an irreducible Markov chain, and two distinct states i and j, such that $f_{ij}^{(n)} > 0$ for all $n \in \mathbf{N}$, and such that $f_{ij}^{(n)}$ is *not* a decreasing function of n (i.e. for some $n \in \mathbf{N}$, $f_{ij}^{(n)} < f_{ij}^{(n+1)}$).

Exercise 8.5.8. Prove the identity $f_{ij} = p_{ij} + \sum_{k \neq j} p_{ik} f_{kj}$. [Hint: Condition on X_1.]

Exercise 8.5.9. For each of the following transition probability matrices, determine which states are recurrent and which are transient, and also compute f_{i1} for each i.

(a) $\begin{pmatrix} 1/2 & 1/2 & 0 & 0 \\ 2/3 & 0 & 0 & 1/3 \\ 0 & 0 & 4/5 & 1/5 \\ 1 & 0 & 0 & 0 \end{pmatrix} .$

(b) $\begin{pmatrix} 1 & 0 & 0 & 0 & 0 & 0 & 0 \\ 1/2 & 0 & 1/2 & 0 & 0 & 0 & 0 \\ 0 & 1/5 & 4/5 & 0 & 0 & 0 & 0 \\ 0 & 0 & 1/3 & 1/3 & 1/3 & 0 & 0 \\ 1/10 & 0 & 0 & 0 & 7/10 & 0 & 1/5 \\ 0 & 0 & 0 & 0 & 0 & 0 & 1 \\ 0 & 0 & 0 & 0 & 0 & 1 & 0 \end{pmatrix} .$

Exercise 8.5.10. Consider a Markov chain (not necessarily irreducible) on a finite state space.

(a) Prove that at least one state must be recurrent.

(b) Give an example where exactly one state is recurrent (and all the rest are transient).

(c) Show by example that if the state space is countably infinite then part (a) is no longer true.

Exercise 8.5.11. For asymmetric one-dimensional simple random walk (i.e. where $\mathbf{P}(X_n = X_{n-1} + 1) = p = 1 - \mathbf{P}(X_n = X_{n-1} - 1)$ for some $p \neq \frac{1}{2}$), provide an asymptotic upper bound for $\sum_{n=N}^{\infty} p_{ii}^{(n)}$. That is, find an explicit expression γ_N, with $\lim_{N \to \infty} \gamma_N = 0$, such that $\sum_{n=N}^{\infty} p_{ii}^{(n)} \leq \gamma_N$ for all sufficiently large N.

Exercise 8.5.12. Let $P = (p_{ij})$ be the matrix of transition probabilities for a Markov chain on a finite state space.

(a) Prove that P always has 1 as an eigenvalue. [Hint: Recall that the eigenvalues of P are the same whether it acts on row vectors to the left or on column vectors to the right.]

(b) Suppose that v is a row eigenvector for P corresponding to the eigenvalue 1, so that $vP = v$. Does v necessarily correspond to a stationary distribution? Why or why not?

Exercise 8.5.13. Call a Markov chain *doubly stochastic* if its transition matrix $\{p_{ij}\}_{i,j \in S}$ has the property that $\sum_{i \in S} p_{ij} = 1$ for each $j \in S$. Prove that, for a doubly stochastic Markov chain on a finite state space S, the uniform distribution (i.e. $\pi_i = 1/|S|$ for each $i \in S$) is a stationary distribution.

Exercise 8.5.14. Give an example of a Markov chain on a finite state space, such that three of the states each have a different period.

Exercise 8.5.15. Consider Ehrenfest's Urn (Example 8.0.4).

(a) Compute $\mathbf{P}_0(X_n = 0)$ for n odd.

(b) Prove that $\mathbf{P}_0(X_n = 0) \not\to 2^{-d}$ as $n \to \infty$.

(c) Why does this not contradict Theorem 8.3.10?

Exercise 8.5.16. Consider the Markov chain with state space $S = \{1, 2, 3\}$ and transition probabilities given by

$$(p_{ij}) = \begin{pmatrix} 0 & 2/3 & 1/3 \\ 1/4 & 0 & 3/4 \\ 4/5 & 1/5 & 0 \end{pmatrix}.$$

(a) Find an explicit formula for $\mathbf{P}_1(\tau_1 = n)$ for each $n \in \mathbf{N}$, where $\tau_1 =$

$\inf\{n \geq 1; \ X_n = 1\}$.
(b) Compute the mean return time $m_1 = \mathbf{E}(\tau_1)$.
(c) Prove that this Markov chain has a unique stationary distribution, to be called $\{\pi_i\}$.
(d) Compute the stationary probability π_1.

Exercise 8.5.17. Give an example of a Markov chain for which some states are positive recurrent, some states are null recurrent, and some states are transient.

Exercise 8.5.18. Prove that if $f_{ij} > 0$ and $f_{ji} = 0$, then i is transient.

Exercise 8.5.19. Prove that for a Markov chain on a *finite* state space, no states are null recurrent. [Hint: The previous exercise provides a starting point.]

Exercise 8.5.20. **(a)** Give an example of a Markov chain on a finite state space which has *multiple* (i.e. two or more) stationary distributions.
(b) Give an example of a *reducible* Markov chain on a finite state space, which nevertheless has a *unique* stationary distribution.
(c) Suppose that a Markov chain on a finite state space is *decomposable*, meaning that the state space can be partitioned as $S = S_1 \mathbin{\dot{\cup}} S_2$, with S_i non-empty, such that $f_{ij} = f_{ji} = 0$ whenever $i \in S_1$ and $j \in S_2$. Prove that the chain has multiple stationary distributions.
(d) Prove that for a Markov chain as in part (b), some states are transient. [Hint: Exercise 8.5.18 may help.]

8.6. Section summary.

This section gave a lengthy exploration of Markov chains (in discrete time and space). It gave several examples of Markov chains, and proved their existence. It defined the important concepts of transience and recurrence, and proved a number of equivalences of them. For an *irreducible* Markov chain, either all states are transient or all states are recurrent.

The section then discussed stationary distributions. It proved that the transition probabilities of an irreducible, aperiodic Markov chain will converge to the stationary distribution, regardless of the starting point. Finally, it related stationary distributions to the mean return times of the chain's states, proving that (for an irreducible chain) a stationary distribution exists if and only if these mean return times are all finite.

9. More probability theorems.

In this section, we discuss a few probability ideas that we have not needed so far, but which will be important for the more advanced material to come.

9.1. Limit theorems.

Suppose $X, X_1, X_2, \ldots$ are random variables defined on some probability triple $(\Omega, \mathcal{F}, \mathbf{P})$. Suppose further that $\lim_{n \to \infty} X_n(\omega) = X(\omega)$ for each fixed $\omega \in \Omega$ (or at least for all ω outside a set of probability 0), i.e. that $\{X_n\}$ converges to X almost surely. Does it necessarily follow that $\lim_{n \to \infty} \mathbf{E}(X_n) = \mathbf{E}(X)$?

We already know that this is not the case in general. Indeed, it was not the case for the "double 'til you win" gambling strategy of Subsection 7.3 (where if $0 < p \leq 1/2$, then $\mathbf{E}(X_n) \leq a$ for all n, even though $\mathbf{E}(\lim X_n) = a + 1$). Or for a simple counter-example, let $\Omega = \mathbf{N}$, with $\mathbf{P}(\omega) = 2^{-\omega}$ for $\omega \in \Omega$, and let $X_n(\omega) = 2^n \delta_{\omega,n}$ (i.e., $X_n(\omega) = 2^n$ if $\omega = n$, and equals 0 otherwise). Then $\{X_n\}$ converges to 0 with probability 1, but $\mathbf{E}(X_n) = 1 \not\to 0$ as $n \to \infty$.

On the other hand, we already have two results giving conditions under which it *is* true that $\mathbf{E}(X_n) \to \mathbf{E}(X)$, namely the Monotone Convergence Theorem (Theorem 4.2.2) and the Bounded Convergence Theorem (Theorem 7.3.1). Such limit theorems are sometimes very helpful.

In this section, we shall establish two more similar limit theorems, namely the Dominated Convergence Theorem and the Uniformly Integrable Convergence Theorem. A first key result is

Theorem 9.1.1.   *(Fatou's Lemma) If $X_n \geq C$ for all n, and some constant $C > -\infty$, then*

$$\mathbf{E}\left(\liminf_{n \to \infty} X_n\right) \leq \liminf_{n \to \infty} \mathbf{E}(X_n).$$

(We allow the possibility that both sides are infinite.)

Proof.   Set $Y_n = \inf_{k \geq n} X_k$, and let $Y = \lim_{n \to \infty} Y_n = \liminf_{n \to \infty} X_n$. Then $Y_n \geq C$ and $\{Y_n\} \nearrow Y$, and furthermore $Y_n \leq X_n$. Hence, by the order-preserving property and the monotone convergence theorem, we have $\liminf_n \mathbf{E}(X_n) \geq \liminf_n \mathbf{E}(Y_n) = \mathbf{E}(Y)$, as claimed. ∎

Remarks.
1. In this theorem, "$\liminf_{n \to \infty} X_n$" is of course interpreted pointwise; that is, its value at ω is $\liminf_{n \to \infty} X_n(\omega)$.

2. For the "simple counter-example" mentioned above, it is easily verified that $\mathbf{E}\left(\liminf_{n\to\infty} X_n\right) = 0$ while $\liminf_{n\to\infty} \mathbf{E}(X_n) = 1$, so the theorem gives $0 \leq 1$ which is true. If we replace X_n by $-X_n$ in that example, we instead obtain $0 \leq -1$ which is false; however, in that case the $\{X_n\}$ are not bounded below.

3. For the "double 'til you win" gambling strategy of Subsection 7.3, the fortunes $\{X_n\}$ are not bounded below. However, they are bounded above (by $a + 1$), so their negatives are bounded below. When the theorem is applied to their negatives, it gives that $-(a + 1) \leq \liminf \mathbf{E}(-X_n)$, so that $\limsup \mathbf{E}(X_n) \leq a + 1$, which is certainly true since $X_n \leq a + 1$ for all n. (In fact, if $p \leq \frac{1}{2}$, then $\limsup \mathbf{E}(X_n) \leq a$.)

It is now straightforward to prove

Theorem 9.1.2.   *(The Dominated Convergence Theorem) If $X, X_1, X_2, \ldots$ are random variables, and if $\{X_n\} \to X$ with probability 1, and if there is a random variable Y with $|X_n| \leq Y$ for all n and with $\mathbf{E}(Y) < \infty$, then $\lim_{n\to\infty} \mathbf{E}(X_n) = \mathbf{E}(X)$.*

Proof.   We note that $Y + X_n \geq 0$. Hence, applying Fatou's Lemma to $\{Y + X_n\}$, we obtain that

$$\mathbf{E}(Y) + \mathbf{E}(X) = \mathbf{E}(Y + X) \leq \liminf_n \mathbf{E}(Y + X_n) = \mathbf{E}(Y) + \liminf_n \mathbf{E}(X_n).$$

Hence, canceling the $\mathbf{E}(Y)$ terms (which is where we use the fact that $\mathbf{E}(Y) < \infty$), we see that $\mathbf{E}(X) \leq \liminf_n \mathbf{E}(X_n)$.

Similarly, $Y - X_n \geq 0$, and applying Fatou's Lemma to $\{Y - X_n\}$, we obtain that $\mathbf{E}(Y) - \mathbf{E}(X) \leq \mathbf{E}(Y) + \liminf_n \mathbf{E}(-X_n) = \mathbf{E}(Y) - \limsup_n \mathbf{E}(X_n)$, so that $\mathbf{E}(X) \geq \limsup_n \mathbf{E}(X_n)$.

But we always have $\limsup_n \mathbf{E}(X_n) \geq \liminf_n \mathbf{E}(X_n)$. Hence, we must have that $\limsup_n \mathbf{E}(X_n) = \liminf_n \mathbf{E}(X_n) = \mathbf{E}(X)$, as claimed. ∎

Remark.   Of course, if the random variable Y is constant, then the dominated convergence theorem reduces to the bounded convergence theorem.

For our second new limit theorem, we need a definition. Note that for any random variable X with $\mathbf{E}(|X|) < \infty$ (i.e., with X "integrable"), by the dominated convergence theorem $\lim_{\alpha\to\infty} \mathbf{E}\left(|X|\mathbf{1}_{|X|\geq\alpha}\right) = \mathbf{E}(0) = 0$. Taking a supremum over n makes a collection of random variables *uniformly integrable*:

Definition 9.1.3. A collection $\{X_n\}$ of random variables is *uniformly integrable* if

$$\lim_{\alpha \to \infty} \sup_n \mathbf{E}\left(|X_n|\mathbf{1}_{|X_n| \geq \alpha}\right) = 0. \qquad (9.1.4)$$

Uniform integrability immediately implies boundedness of certain expectations:

Proposition 9.1.5. *If $\{X_n\}$ is uniformly integrable, then $\sup_n \mathbf{E}(|X_n|) < \infty$. Furthermore, if also $\{X_n\} \to X$ a.s., then $\mathbf{E}|X| < \infty$.*

Proof. Choose α_1 so that (say) $\sup_n \mathbf{E}\left(|X_n|\mathbf{1}_{|X_n| \geq \alpha_1}\right) \leq 1$. Then

$$\sup_n \mathbf{E}\left(|X_n|\right) = \sup_n \mathbf{E}\left(|X_n|\mathbf{1}_{|X_n| < \alpha_1} + |X_n|\mathbf{1}_{|X_n| \geq \alpha_1}\right) \leq \alpha_1 + 1 < \infty.$$

It then follows from Fatou's Lemma that, if $\{X_n\} \to X$ a.s., then $\mathbf{E}(|X|) \leq \liminf_n \mathbf{E}(|X_n|) \leq \sup_n \mathbf{E}(|X_n|) < \infty.$ ∎

The main use of uniform integrability is given by:

Theorem 9.1.6. *(The Uniform Integrability Convergence Theorem) If $X, X_1, X_2, \ldots$ are random variables, and if $\{X_n\} \to X$ with probability 1, and if $\{X_n\}$ are uniformly integrable, then $\lim_{n \to \infty} \mathbf{E}(X_n) = \mathbf{E}(X)$.*

Proof. Let $Y_n = |X_n - X|$, so that $Y_n \to 0$. We shall show that $\mathbf{E}(Y_n) \to 0$; it then follows from the triangle inequality that $|\mathbf{E}(X_n) - \mathbf{E}(X)| \leq \mathbf{E}(Y_n) \to 0$, thus proving the theorem. We will consider $\mathbf{E}(Y_n)$ in two pieces, using that $Y_n = Y_n \mathbf{1}_{Y_n < \alpha} + Y_n \mathbf{1}_{Y_n \geq \alpha}$.

Let $Y_n^{(\alpha)} = Y_n \mathbf{1}_{Y_n < \alpha}$. Then for any fixed $\alpha > 0$, we have $|Y_n^{(\alpha)}| \leq \alpha$, and also $Y_n^{(\alpha)} \to 0$ as $n \to \infty$, so by the bounded convergence theorem we have

$$\lim_{n \to \infty} \mathbf{E}\left(Y_n^{(\alpha)}\right) = 0, \qquad \alpha > 0. \qquad (9.1.7)$$

For the second piece, we note by the triangle inequality (again) that $Y_n \leq |X_n| + |X| \leq 2M_n$ where $M_n = \max(|X_n|, |X|)$. Hence, if $Y_n \geq \alpha$, then we must have $M_n \geq \alpha/2$, and thus that either $|X_n| \geq \alpha/2$ or $|X| \geq \alpha/2$. This implies that

$$Y_n\mathbf{1}_{Y_n \geq \alpha} \leq 2M_n\mathbf{1}_{M_n \geq \alpha/2} \leq 2|X_n|\mathbf{1}_{|X_n| \geq \alpha/2} + 2|X|\mathbf{1}_{|X| \geq \alpha/2}.$$

Taking expectations and supremums gives

$$\sup_n \mathbf{E}\left(Y_n\mathbf{1}_{Y_n \geq \alpha}\right) \leq 2\sup_n \mathbf{E}\left(|X_n|\mathbf{1}_{|X_n| \geq \alpha/2}\right) + 2\mathbf{E}\left(|X|\mathbf{1}_{|X| \geq \alpha/2}\right).$$

Now, as $\alpha \to \infty$, the first of these two terms goes to 0 by the uniform integrability assumption; and the second goes to 0 by the dominated convergence theorem (since $\mathbf{E}|X| < \infty$). We conclude that

$$\lim_{\alpha \to \infty} \sup_n \mathbf{E}\left(Y_n 1_{Y_n \geq \alpha}\right) = 0. \qquad (9.1.8)$$

To finish the proof, let $\epsilon > 0$. By (9.1.8), we can find $\alpha_0 > 0$ such that $\sup_n \mathbf{E}\left(Y_n 1_{Y_n \geq \alpha_0}\right) < \epsilon/2$. By (9.1.7), we can find $n_0 \in \mathbf{N}$ such that $\mathbf{E}\left(Y_n^{(\alpha_0)}\right) < \epsilon/2$ for all $n \geq n_0$. Then, for any $n \geq n_0$, we have

$$\mathbf{E}(Y_n) = \mathbf{E}\left(Y_n^{(\alpha_0)}\right) + \mathbf{E}\left(Y_n 1_{Y_n \geq \alpha_0}\right) < \frac{\epsilon}{2} + \frac{\epsilon}{2} = \epsilon.$$

Hence, $\mathbf{E}(Y_n) \to 0$, as desired.   ∎

Remark 9.1.9.   While the above limit theorems were all stated for a *countable* limit of random variables, they apply equally well for continuous limits. For example, suppose $\{X_t\}_{t \geq 0}$ is a continuous-parameter family of random variables, and that $\lim_{t \searrow 0} X_t(\omega) = X_0(\omega)$ for each fixed $\omega \in \Omega$. Suppose further that the family $\{X_t\}_{t \geq 0}$ is dominated (or uniformly integrable) as in the previous limit theorems. Then for any countable sequence of parameters $\{t_n\} \searrow 0$, the theorems say that $\mathbf{E}(X_{t_n}) \to \mathbf{E}(X_0)$. Since this is true for *any* sequence $\{t_n\} \searrow 0$, it follows that $\lim_{t \searrow 0} \mathbf{E}(X_t) = \mathbf{E}(X_0)$ as well.

9.2. Differentiation of expectation.

A classic question from multivariable calculus asks when we can "differentiate under the integral sign". For example, is it true that

$$\frac{d}{dt} \int_0^1 e^{st} ds = \int_0^1 \left[\frac{\partial}{\partial t} e^{st}\right] ds = \int_0^1 s e^{st} ds \,?$$

The answer is yes, as the following proposition shows. More generally, the proposition considers a family of random variables $\{F_t\}$ (e.g. $F_t(\omega) = e^{\omega t}$), and the derivative (with respect to t) of the function $\mathbf{E}(F_t)$ (e.g. of $\int_0^1 e^{\omega t} d\omega$).

Proposition 9.2.1.   Let $\{F_t\}_{a < t < b}$ be a collection of random variables with finite expectations, defined on some probability triple $(\Omega, \mathcal{F}, \mathbf{P})$. Suppose for each ω and each $a < t < b$, the derivative $F_t'(\omega) = \frac{\partial}{\partial t} F_t(\omega)$ exists. Then F_t' is a random variable. Suppose further that there is a random variable Y on $(\Omega, \mathcal{F}, \mathbf{P})$ with $\mathbf{E}(Y) < \infty$, such that $|F_t'| \leq Y$ for all $a < t < b$.

Then if we define $\phi(t) = \mathbf{E}(F_t)$, then ϕ is differentiable, with finite derivative $\phi'(t) = \mathbf{E}(F_t')$ for all $a < t < b$.

Proof. To see that F_t' is a random variable, let $t_n = t + \frac{1}{n}$. Then $F_t' = \lim_{n \to \infty} n(F_{t_n} - F_t)$ and hence is the countable limit of random variables, and therefore is itself a random variable by (3.1.6).

Next, note that we always have $\left| \frac{F_{t+h} - F_t}{h} \right| \leq Y$. Hence, using the dominated convergence theorem together with Remark 9.1.9, we have

$$\phi'(t) = \lim_{h \to 0} \frac{\phi(t + h) - \phi(t)}{h} = \lim_{h \to 0} \mathbf{E}\left(\frac{F_{t+h} - F_t}{h} \right)$$

$$= \mathbf{E}\left(\lim_{h \to 0} \frac{F_{t+h} - F_t}{h} \right) = \mathbf{E}(F_t'),$$

and this is finite since $\mathbf{E}(|F_t'|) \leq \mathbf{E}(Y) < \infty$. ∎

9.3. Moment generating functions and large deviations.

The *moment generating function* of a random variable X is the function

$$M_X(s) = \mathbf{E}(e^{sX}), \qquad s \in \mathbf{R}.$$

At first glance, it may appear that this function is of little use. However, we shall see that a surprising amount of information about the distribution of X can be obtained from $M_X(s)$.

If X and Y are independent random variables, then e^{sX} and e^{sY} are independent by Proposition 3.2.3, so we have by (4.2.7) that

$$M_{X+Y}(s) = M_X(s) M_Y(s), \qquad X, Y \text{ independent}. \qquad (9.3.1)$$

Clearly, we always have $M_X(0) = 1$. However, we may have $M_X(s) = \infty$ for certain $s \neq 0$. For example, if $\mathbf{P}[X = m] = c/m^2$ for all integers $m \neq 0$, where $c = (\sum_{j \neq 0} j^{-2})^{-1}$, then $M_X(s) = \infty$ for *all* $s \neq 0$. On the other hand, if $X \sim N(0, 1)$, then completing the square gives that

$$M_X(s) = \int_{-\infty}^{\infty} e^{sx} \frac{1}{\sqrt{2\pi}} e^{-x^2/2} dx = e^{s^2/2} \int_{-\infty}^{\infty} \frac{1}{\sqrt{2\pi}} e^{-(x-s)^2/2} dx$$

$$= e^{s^2/2}(1) = e^{s^2/2}. \qquad (9.3.2)$$

A key property of moment generating functions, at least when they are finite in a neighbourhood of 0, is the following.

Theorem 9.3.3.   *Let X be a random variable such that $M_X(s) < \infty$ for $|s| < s_0$, for some $s_0 > 0$. Then $\mathbf{E}|X^n| < \infty$ for all n, and $M_X(s)$ is analytic for $|s| < s_0$ with*

$$M_X(s) = \sum_{n=0}^{\infty} \mathbf{E}(X^n)s^n/n!.$$

In particular, the rth derivative at $s = 0$ is given by $M_X^{(r)}(0) = \mathbf{E}(X^r)$.

Proof.   The idea of the proof is that

$$M_X(s) = \mathbf{E}(e^{sX}) = \mathbf{E}\left(1 + (sX) + (sX)^2/2! + \ldots\right)$$

$$= 1 + s\mathbf{E}(X) + \frac{s^2}{2!}\mathbf{E}(X^2) + \ldots.$$

However, the final equality requires justification. For this, fix s with $|s| < s_0$, and let $Z_n = 1 + (sX) + (sX)^2/2! + \ldots + (sX)^n/n!$. We have to show that $\mathbf{E}(\lim_{n\to\infty} Z_n) = \lim_{n\to\infty} \mathbf{E}(Z_n)$. Now, for all $n \in \mathbf{N}$,

$$|Z_n| \leq 1 + |sX| + |sX|^2/2! + \ldots + |sX|^n/n!$$

$$\leq 1 + |sX| + |sX|^2/2! + \ldots = e^{|sX|} \leq e^{sX} + e^{-sX} \equiv Y.$$

Since $|s| < s_0$, we have $\mathbf{E}(Y) = M_X(s) + M_X(-s) < \infty$. Hence, by the dominated convergence theorem, $\mathbf{E}(\lim_{n\to\infty} Z_n) = \lim_{n\to\infty} \mathbf{E}(Z_n)$. ∎

Remark.   Theorem 9.3.3 says that the rth derivative of M_X at 0 equals the rth moment of X (thus explaining the terminology "moment generating function"). For example, $M_X(0) = 1$, $M_X'(0) = \mathbf{E}(X)$, $M_X''(0) = \mathbf{E}(X^2)$, etc. This result also follows since $(\frac{d}{ds})^r \mathbf{E}(e^{sX}) = \mathbf{E}\left((\frac{\partial}{\partial s})^r e^{sX}\right) = \mathbf{E}\left(X^r e^{sX}\right)$, where the exchange of derivative and expectation can be justified (for $|s| < s_0$) using Proposition 9.2.1 and induction.

We now consider the subject of *large deviations*. If $X_1, X_2, \ldots$ are i.i.d. with common mean m and finite variance v, then it follows from Chebychev's inequality (as in the proof of the Weak Law of Large Numbers) that for all $\epsilon > 0$, $\mathbf{P}\left(\frac{X_1 + \ldots + X_n}{n} \geq m + \epsilon\right) \leq v/n\epsilon^2$, which converges to 0 as $n \to \infty$. But how quickly is this limit reached? Does the probability really just decrease as $O(1/n)$, or does it decrease faster? In fact, if the moment generating functions are finite in a neighbourhood of 0, then the convergence is exponentially fast:

Theorem 9.3.4.   *Suppose $X_1, X_2, \ldots$ are i.i.d. with common mean m, and with $M_{X_i}(s) < \infty$ for $-a < s < b$ where $a, b > 0$. Then*

$$\mathbf{P}\left(\frac{X_1 + \ldots + X_n}{n} \geq m + \epsilon\right) \leq \rho^n, \qquad n \in \mathbf{N},$$

where $\rho = \inf_{0 < s < b} \left(e^{-s(m+\epsilon)} M_{X_1}(s) \right) < 1$.

This theorem gives an exponentially small upper bound on the probability that the average of the X_i exceeds its mean by at least ϵ. This is a (very simple) example of a *large deviations* result, and shows that in this case the probability is decreasing to zero exponentially quickly – much faster than just $O(1/n)$.

To prove Theorem 9.3.4, we begin with a lemma:

Lemma 9.3.5. Let Z be a random variable with $\mathbf{E}(Z) < 0$, such that $M_Z(s) < \infty$ for $-a < s < b$, for some $a, b > 0$. Then $\mathbf{P}(Z \geq 0) \leq \rho$, where $\rho = \inf_{0 < s < b} M_Z(s) < 1$.

Proof. For any $0 < s < b$, since the function $x \mapsto e^{sx}$ is increasing, Markov's inequality implies that

$$\mathbf{P}(Z \geq 0) = \mathbf{P}(e^{sZ} \geq 1) \leq \frac{\mathbf{E}(e^{sZ})}{1} = M_Z(s).$$

Hence, taking the infimum over $0 < s < b$,

$$\mathbf{P}(Z \geq 0) \leq \inf_{0 < s < b} M_Z(s) = \rho.$$

Furthermore, since $M_Z(0) = 1$ and $M_Z'(0) = \mathbf{E}(Z) < 0$, we must have that $M_Z(s) < 1$ for all positive s sufficiently close to 0. In particular, $\rho < 1$. ■

Remark. In Lemma 9.3.5, we need $a > 0$ to ensure that $M_Z'(0) = \mathbf{E}(Z)$. However, the precise value of a is unimportant.

Proof of Theorem 9.3.4. Let $Y_i = X_i - m - \epsilon$, so $\mathbf{E}(Y_i) = -\epsilon < 0$. Then for $-a < s < b$, $M_{Y_i}(s) = \mathbf{E}(e^{sY_i}) = e^{-s(m+\epsilon)} \mathbf{E}(e^{sX_i}) = e^{-s(m+\epsilon)} M_{X_i}(s) < \infty$. Using Lemma 9.3.5 and (9.3.1), we have

$$\mathbf{P}\left(\frac{X_1 + \ldots + X_n}{n} \geq m + \epsilon \right) = \mathbf{P}\left(\frac{Y_1 + \ldots + Y_n}{n} \geq 0 \right)$$

$$= \mathbf{P}\left(Y_1 + \ldots + Y_n \geq 0 \right) \leq \inf_{0 < s < b} M_{Y_1 + \ldots + Y_n}(s) = \inf_{0 < s < b} \left(M_{Y_1}(s) \right)^n = \rho^n,$$

where $\rho = \inf_{0 < s < b} M_{Y_1}(s) = \inf_{0 < s < b}(e^{-s(m+\epsilon)} M_{X_1}(s))$. Furthermore, from Lemma 9.3.5, $\rho < 1$. ■

9.4. Fubini's Theorem and convolution.

In multivariable calculus, an iterated integral like $\int_0^1 \int_0^1 x^2 y^3 \, dx \, dy$, can be computed in three different ways: integrate first x and then y, or integrate first y and then x, or compute a two-dimensional "double integral" over the full two-dimensional region. It is well known that under mild conditions, these three different integrals will all be equal.

A generalisation of this is Fubini's Theorem, which allows us to compute expectations with respect to product measure in terms of an "iterated integral", where we integrate first with respect to one variable and then with respect to the other, in either order. (For non-negative f, the theorem is sometimes referred to as *Tonelli's theorem*.)

Theorem 9.4.1. *(Fubini's Theorem) Let μ be a probability measure on $\mathcal{X}$, and ν a probability measure on $\mathcal{Y}$, and let $\mu \times \nu$ be product measure on $\mathcal{X} \times \mathcal{Y}$. If $f : \mathcal{X} \times \mathcal{Y} \to \mathbf{R}$ is measurable with respect to $\mu \times \nu$, then*

$$
\begin{aligned}
\int_{\mathcal{X} \times \mathcal{Y}} f \, d(\mu \times \nu) &= \int_{\mathcal{X}} \left(\int_{\mathcal{Y}} f(x,y) \nu(dy) \right) \mu(dx) \\
&= \int_{\mathcal{Y}} \left(\int_{\mathcal{X}} f(x,y) \mu(dx) \right) \nu(dy),
\end{aligned} \tag{9.4.2}
$$

provided that either $\int_{\mathcal{X} \times \mathcal{Y}} f^+ d(\mu \times \nu) < \infty$ or $\int_{\mathcal{X} \times \mathcal{Y}} f^- d(\mu \times \nu) < \infty$ (or both). [This is guaranteed if, for example, $f \geq C > -\infty$, or $f \leq C < \infty$, or $\int_{\mathcal{X} \times \mathcal{Y}} |f| \, d(\mu \times \nu) < \infty$. Note that we allow that the inner integrals (i.e., the integrals inside the brackets) may be infinite or even undefined on a set of probability 0.]

The proof of Theorem 9.4.1 requires one technical lemma:

Lemma 9.4.3. *The mapping $E \mapsto \int_{\mathcal{Y}} \left(\int_{\mathcal{X}} 1_E(x,y) \mu(dx) \right) \nu(dy)$ is a well-defined, countably additive function of subsets E.*

Proof (optional). For $E \subseteq \mathcal{X} \times \mathcal{Y}$, and $y \in \mathcal{Y}$, let $S_y(E) = \{x \in \mathcal{X}; (x,y) \in E\}$. We first argue that, for any $y \in \mathcal{Y}$ and any set E which is measurable with respect to $\mu \times \nu$, the set $S_y(E)$ is measurable with respect to μ. Indeed, this is certainly true if $E = A \times B$ with A and B measurable, for then $S_y(E)$ is always either A or $\emptyset$. On the other hand, since S_y preserves set operations (i.e., $S_y(E_1 \cup E_2) = S_y(E_1) \cup S_y(E_2)$, $S_y(E^C) = S_y(E)^C$, etc.), the collection of sets E for which $S_y(E)$ is measurable is a σ-algebra. Furthermore, the measurable rectangles $A \times B$ generate the product measure's entire σ-algebra. Hence, for any measurable set E, $S_y(E)$ is a measurable subset of $\mathcal{X}$, so that $\mu(S_y(E))$ is well-defined.

Next, consider the collection $\mathcal{G}$ of those measurable subsets $E \subseteq \mathcal{X} \times \mathcal{Y}$ for which $\mu(S_y(E))$ is a measurable function of $y \in \mathcal{Y}$. This $\mathcal{G}$ certainly contains all the measurable rectangles $A \times B$, since then $\mu(S_y(E)) =$

$\mu(A)\,\mathbf{1}_B(y)$ which is measurable. Also, $\mathcal{G}$ is closed under complements since $\mu\big(S_y(E^C)\big) = \mu\big(S_y(E)^C\big) = 1 - \mu\big(S_y(E)\big)$. It is also closed under countable disjoint unions, since if $\{E_n\}$ are disjoint, then so are $\{S_y(E_n)\}$, and thus $\mu\big(S_y(\bigcup_n E_n)\big) = \mu\big(\bigcup_n S_y(E_n)\big) = \sum_n \mu\big(S_y(E_n)\big)$. From these facts, it follows (formally, from the "π-λ theorem", e.g. Billingsley, 1995, p. 42) that $\mathcal{G}$ includes all measurable subsets E, i.e. that $\mu\big(S_y(E)\big)$ is a measurable function of y for any measurable set E. Thus, integrals like $\int_{\mathcal{Y}} \mu\big(S_y(E)\big)\,\nu(dy)$ are well-defined. Since $\int_{\mathcal{Y}} \big(\int_{\mathcal{X}} \mathbf{1}_E(x,y)\mu(dx)\big)\,\nu(dy) = \int_{\mathcal{Y}} \mu\big(S_y(E)\big)\,\nu(dy)$, the claim about being well-defined follows.

To prove countable additivity, let $\{E_n\}$ be disjoint. Then, as above, $\mu\big(S_y(\bigcup_n E_n)\big) = \sum_n \mu\big(S_y(E_n)\big)$. Countable additivity then follows from countable linearity of expectations with respect to ν. ∎

Proof of Theorem 9.4.1. We first consider the case where $f = \mathbf{1}_E$ is an indicator function of a measurable set $E \subseteq \mathcal{X} \times \mathcal{Y}$. By Lemma 9.4.3, the mapping $E \mapsto \int_{\mathcal{Y}} \big(\int_{\mathcal{X}} \mathbf{1}_E(x,y)\mu(dx)\big)\,\nu(dy)$ is a well-defined, countably additive function of subsets E. When $E = A \times B$, this integral clearly equals $\mu(A)\,\nu(B) = (\mu \times \nu)(E)$. Hence, by uniqueness of extensions (Proposition 2.5.8), we must have $\int_{\mathcal{Y}} \big(\int_{\mathcal{X}} \mathbf{1}_E(x,y)\mu(dx)\big)\,\nu(dy) = (\mu \times \nu)(E)$ for any measurable subset E. Similarly, $\int_{\mathcal{X}} \big(\int_{\mathcal{Y}} \mathbf{1}_E(x,y)\nu(dy)\big)\,\mu(dx) = (\mu \times \nu)(E)$, so that (9.4.2) holds whenever $f = \mathbf{1}_E$.

We complete the proof by our usual arguments. Indeed, by linearity, (9.4.2) holds whenever f is a simple function. Then, by the monotone convergence theorem, (9.4.2) holds for general measurable non-negative f. Finally, again by linearity since $f = f^+ - f^-$, we see that (9.4.2) holds for general functions f as long as we avoid the $\infty - \infty$ case where $\int_{\mathcal{X} \times \mathcal{Y}} f^+ d(\mu \times \nu) = \int_{\mathcal{X} \times \mathcal{Y}} f^- d(\mu \times \nu) = \infty$ (while still allowing that the inner integrals may be $\infty - \infty$ on a set of probability 0). ∎

Remark. If $\int_{\mathcal{X} \times \mathcal{Y}} f^+ d(\mu \times \nu) = \int_{\mathcal{X} \times \mathcal{Y}} f^- d(\mu \times \nu) = \infty$, then Fubini's Theorem might not hold; see Exercises 9.5.13 and 9.5.14.

By additivity, Fubini's Theorem still holds if μ and/or ν are σ-finite measures (cf. Remark 4.4.3), rather than just probability measures. In particular, letting $\mu = \mathbf{P}$ and ν be counting measure on $\mathbf{N}$ leads to the following generalisation of countable linearity (essentially a re-statement of Exercise 4.5.14(a)):

Corollary 9.4.4. Let $Z_1, Z_2, \ldots$ be random variables with $\sum_i \mathbf{E}|Z_i| < \infty$. Then $\mathbf{E}(\sum_i Z_i) = \sum_i \mathbf{E}(Z_i)$.

As an application of Fubini's Theorem, we consider the *convolution* formula, about sums of independent random variables.

Theorem 9.4.5. *Suppose X and Y are independent random variables with distributions $\mu = \mathcal{L}(X)$ and $\nu = \mathcal{L}(Y)$. Then the distribution of $X+Y$ is given by $\mu * \nu$, where*

$$(\mu * \nu)(H) = \int_{\mathbf{R}} \mu(H - y)\, \nu(dy), \qquad H \subseteq \mathbf{R},$$

with $H - y = \{h - y;\ h \in H\}$. Furthermore, if μ has density f and ν has density g (with respect to $\lambda = $ Lebesgue measure on $\mathbf{R}$), then $\mu * \nu$ has density $f * g$, where

$$(f * g)(x) = \int_{\mathbf{R}} f(x - y)\, g(y)\, \lambda(dy), \qquad x \in \mathbf{R}.$$

Proof. Since X and Y are independent, we know that $\mathcal{L}((X, Y)) = \mu \times \nu$, i.e. the distribution of the ordered pair (X, Y) is equal to product measure. Given a Borel subset $H \subseteq \mathbf{R}$, let $B = \{(x, y) \in \mathbf{R}^2;\ x + y \in H\}$. Then using Fubini's Theorem, we have

$$
\begin{aligned}
\mathbf{P}(X + Y \in H) &= \mathbf{P}((X, Y) \in B) \\
&= (\mu \times \nu)(B) \\
&= \int_{\mathbf{R} \times \mathbf{R}} \mathbf{1}_B \, d(\mu \times \nu) \\
&= \int_{\mathbf{R}} \left(\int_{\mathbf{R}} \mathbf{1}_B(x, y)\, \mu(dx) \right) \nu(dy) \\
&= \int_{\mathbf{R}} \mu\{x \in \mathbf{R};\ (x, y) \in B\}\, \nu(dy) \\
&= \int_{\mathbf{R}} \mu(H - y)\, \nu(dy) \\
&\equiv (\mu * \nu)(H),
\end{aligned}
$$

so $\mathcal{L}(X + Y) = \mu * \nu$. If μ has density f and ν has density g, then using Proposition 6.2.3, shift invariance of Lebesgue measure as in (1.2.5), and Fubini's theorem again,

$$
\begin{aligned}
(\mu * \nu)(H) &= \int_{\mathbf{R}} \mu(H - y)\, \nu(dy) \\
&= \int_{\mathbf{R}} \left(\int_{H-y} 1\, \mu(dx) \right) \nu(dy) \\
&= \int_{\mathbf{R}} \left(\int_{H-y} f(x)\, \lambda(dx) \right) g(y) \lambda(dy) \\
&= \int_{\mathbf{R}} \left(\int_{H} f(x - y)\, \lambda(dx) \right) g(y)\, \lambda(dy) \\
&= \int_{\mathbf{R}} \left(\int_{H} f(x - y)\, g(y)\, \lambda(dx) \right) \lambda(dy) \\
&= \int_{H} \left(\int_{\mathbf{R}} f(x - y)\, g(y)\, \lambda(dy) \right) \lambda(dx) \\
&\equiv \int_{H} (f * g)(x)\, \lambda(dx),
\end{aligned}
$$

so $\mu * \nu$ has density given by $f * g$. ∎

9.5. Exercises.

Exercise 9.5.1. For the "simple counter-example" with $\Omega = \mathbf{N}$, $\mathbf{P}(\omega) = 2^{-\omega}$ for $\omega \in \Omega$, and $X_n(\omega) = 2^n \, \delta_{\omega,n}$, verify explicitly that the hypotheses of each of the Monotone Convergence Theorem, the Bounded Convergence Theorem, the Dominated Convergence Theorem, and the Uniform Integrability Convergence Theorem, are all violated.

Exercise 9.5.2. Give an example of a sequence of random variables which is unbounded but still uniformly integrable. For bonus points, make the sequence also be undominated, i.e. violate the hypothesis of the Dominated Convergence Theorem.

Exercise 9.5.3. Let $X, X_1, X_2, \ldots$ be non-negative random variables, defined jointly on some probability triple $(\Omega, \mathcal{F}, \mathbf{P})$, each having finite expected value. Assume that $\lim_{n\to\infty} X_n(\omega) = X(\omega)$ for all $\omega \in \Omega$. For $n, K \in \mathbf{N}$, let $Y_{n,K} = \min(X_n, K)$. For each of the following statements, either prove it must true, or provide a counter-example to show it is sometimes false.
(a) $\lim_{K\to\infty} \lim_{n\to\infty} \mathbf{E}(Y_{n,K}) = \mathbf{E}(X)$.
(b) $\lim_{n\to\infty} \lim_{K\to\infty} \mathbf{E}(Y_{n,K}) = \mathbf{E}(X)$.

Exercise 9.5.4. Suppose that $\lim_{n\to\infty} X_n(\omega) = 0$ for all $\omega \in \Omega$, but $\lim_{n\to\infty} \mathbf{E}[X_n] \neq 0$. Prove that $\mathbf{E}(\sup_n |X_n|) = \infty$.

Exercise 9.5.5. Suppose $\sup_n \mathbf{E}(|X_n|^r) < \infty$ for some $r > 1$. Prove that $\{X_n\}$ is uniformly integrable. [Hint: If $|X_n(\omega)| \geq \alpha > 0$, then $|X_n(\omega)| \leq |X_n(\omega)|^r / \alpha^{r-1}$.]

Exercise 9.5.6. Prove that Theorem 9.1.6 implies Theorem 9.1.2. [Hint: Suppose $|X_n| \leq Y$ where $\mathbf{E}(Y) < \infty$. Prove that $\{X_n\}$ satisfies (9.1.4).]

Exercise 9.5.7. Prove that Theorem 9.1.2 implies Theorem 4.2.2, assuming that $\mathbf{E}|X| < \infty$. [Hint: Suppose $\{X_n\} \nearrow X$ where $\mathbf{E}|X| < \infty$. Prove that $\{X_n\}$ is dominated.]

Exercise 9.5.8. Let $\Omega = \{1, 2\}$, with $\mathbf{P}(\{1\}) = \mathbf{P}(\{2\}) = \frac{1}{2}$, and let $F_t(\{1\}) = t^2$ and $F_t(\{2\}) = t^4$ for $0 < t < 1$.
(a) What does Proposition 9.2.1 conclude in this case?
(b) In light of the above, what rule from calculus is implied by Proposition 9.2.1?

Exercise 9.5.9. Let $X_1, X_2, \ldots$ be i.i.d., each with $\mathbf{P}(X_i = 1) = \mathbf{P}(X_i = -1) = 1/2$.

(a) Compute the moment generating functions $M_{X_i}(s)$.
(b) Use Theorem 9.3.4 to obtain an exponentially-decreasing upper bound on $\mathbf{P}\left(\frac{1}{n}(X_1 + \ldots + X_n) \geq 0.1\right)$.

Exercise 9.5.10.   Let $X_1, X_2, \ldots$ be i.i.d., each having the standard normal distribution $N(0, 1)$. Use Theorem 9.3.4 to obtain an exponentially-decreasing upper bound on $\mathbf{P}\left(\frac{1}{n}(X_1 + \ldots + X_n) \geq 0.1\right)$. [Hint: Don't forget (9.3.2).]

Exercise 9.5.11.   Let X have the distribution Exponential(5), with density $f_X(x) = 5\,e^{-5x}$ for $x \geq 0$ (with $f_X(x) = 0$ for $x < 0$).
(a) Compute the moment generating function $M_X(t)$.
(b) Use $M_X(t)$ to compute (with explanation) the expected value $\mathbf{E}(X)$.

Exercise 9.5.12.   Let $\alpha > 2$, and let $M(t) = e^{-|t|^\alpha}$ for $t \in \mathbf{R}$. Prove that $M(t)$ is not a moment generating function of any probability distribution. [Hint: Consider $M''(t)$.]

Exercise 9.5.13.   Let $\mathcal{X} = \mathcal{Y} = \mathbf{N}$, and let $\mu\{n\} = \nu\{n\} = 2^{-n}$ for $n \in \mathbf{N}$. Let $f : \mathcal{X} \times \mathcal{Y} \to \mathbf{R}$ by $f(n, n) = (4^n - 1)$, and $f(n, n+1) = -2(4^n - 1)$, with $f(n, m) = 0$ otherwise.
(a) Compute $\int_{\mathcal{X}} \left(\int_{\mathcal{Y}} f(x, y)\nu(dy)\right)\mu(dx)$.
(b) Compute $\int_{\mathcal{Y}} \left(\int_{\mathcal{X}} f(x, y)\mu(dx)\right)\nu(dy)$.
(c) Why does the result not contradict Fubini's Theorem?

Exercise 9.5.14.   Let λ be Lebesgue measure on $[0, 1]$, and let $f(x, y) = 8xy(x^2 - y^2)(x^2 + y^2)^{-3}$ for $(x, y) \neq (0, 0)$, with $f(0, 0) = 0$.
(a) Compute $\int_0^1 \left(\int_0^1 f(x, y)\lambda(dy)\right)\lambda(dx)$. [Hint: Make the substitution $u = x^2 + y^2$, $v = x$, so $du = 2\,y\,dy$, $dv = dx$, and $x^2 - y^2 = 2v^2 - u$.]
(b) Compute $\int_0^1 \left(\int_0^1 f(x, y)\lambda(dx)\right)\lambda(dy)$.
(c) Why does the result not contradict Fubini's Theorem?

Exercise 9.5.15.   Let $X \sim \mathbf{Poisson}(a)$ and $Y \sim \mathbf{Poisson}(b)$ be independent. Let $Z = X + Y$. Use the convolution formula to compute $\mathbf{P}(Z = z)$ for all $z \in \mathbf{R}$, and prove that $Z \sim \mathbf{Poisson}(a + b)$.

Exercise 9.5.16.   Let $X \sim N(a, v)$ and $Y \sim N(b, w)$ be independent. Let $Z = X + Y$. Use the convolution formula to prove that $Z \sim N(a+b, v+w)$.

Exercise 9.5.17.   For $\alpha, \beta > 0$, the **Gamma**(α, β) distribution has density function $f(x) = \beta^\alpha x^{\alpha-1} e^{-x/\beta} / \Gamma(\alpha)$ for $x > 0$ (with $f(x) = 0$ for

$x \le 0$), where $\Gamma(\alpha) = \int_0^\infty t^{x-1}e^{-t}dt$ is the *gamma function*. (Hence, when $\alpha = 1$, $\mathbf{Gamma}(1, \beta) = \mathbf{Exp}(\beta)$.) Let $X \sim \mathbf{Gamma}(\alpha, \beta)$ and $Y \sim \mathbf{Gamma}(\gamma, \beta)$ be independent, and let $Z = X + Y$. Use the convolution formula to prove that $Z \sim \mathbf{Gamma}(\alpha + \gamma, \beta)$. [Note: You may use the facts that $\Gamma(\alpha + 1) = \alpha \Gamma(\alpha)$ for $\alpha \in \mathbf{R}$, and $\Gamma(n) = (n-1)!$ for $n \in \mathbf{N}$, and $\int_0^x t^{r-1}(x-t)^{s-1}dt = x^{r+s-1}\Gamma(r)\Gamma(s)/\Gamma(r+s)$ for $r, s, x > 0$.]

9.6. Section summary.

This section presented various probability results that will be required for the more advanced portions of the text. First, the Dominated Convergence Theorem and the Uniform Integrability Limit Theorem were proved, to extend the Monotone Convergence Theorem and the Bounded Convergence Theorem studied previously, providing further conditions under which $\lim \mathbf{E}(X_n) = \mathbf{E}(\lim X_n)$. Second, a result was given allowing derivative and expectation operators to be exchanged. Third, moment generating functions were introduced, and some of their basic properties were studied. Finally, Fubini's Theorem for iterated integration was proved, and applied to give a convolution formula for the distribution of a sum of independent random variables.

10. Weak convergence.

Given Borel probability distributions $\mu, \mu_1, \mu_2, \ldots$ on $\mathbf{R}$, we shall write $\mu_n \Rightarrow \mu$, and say that $\{\mu_n\}$ *converges weakly* to μ, if $\int_{\mathbf{R}} f \, d\mu_n \to \int_{\mathbf{R}} f \, d\mu$ for all bounded continuous functions $f : \mathbf{R} \to \mathbf{R}$.

This is a rather natural[*] definition, though we draw the reader's attention to the fact that this convergence need hold only for *continuous* functions f (as opposed to all Borel-measurable f; cf. Proposition 3.1.8). That is, the "topology" of $\mathbf{R}$ is being used here, not just its measure-theoretic properties.

10.1. Equivalences of weak convergence.

We now present a number of equivalences of weak convergence (see also Exercise 10.3.8). For condition (2), recall that the *boundary* of a set $A \subseteq \mathbf{R}$ is $\partial A = \{x \in \mathbf{R}; \ \forall \epsilon > 0, \ A \cap (x - \epsilon, x + \epsilon) \neq \emptyset, \ A^C \cap (x - \epsilon, x + \epsilon) \neq \emptyset\}$.

Theorem 10.1.1. *The following are equivalent.*
(1) $\mu_n \Rightarrow \mu$ (i.e., $\{\mu_n\}$ converges weakly to μ);
(2) $\mu_n(A) \to \mu(A)$ for all measurable sets A such that $\mu(\partial A) = 0$;
(3) $\mu_n((-\infty, x]) \to \mu((-\infty, x])$ for all $x \in \mathbf{R}$ such that $\mu\{x\} = 0$;
(4) (Skorohod's Theorem) there are random variables $Y, Y_1, Y_2, \ldots$ defined jointly on some probability triple, with $\mathcal{L}(Y) = \mu$ and $\mathcal{L}(Y_n) = \mu_n$ for each $n \in \mathbf{N}$, such that $Y_n \to Y$ with probability 1.
(5) $\int_{\mathbf{R}} f \, d\mu_n \to \int_{\mathbf{R}} f \, d\mu$ for all bounded Borel-measurable functions $f : \mathbf{R} \to \mathbf{R}$ such that $\mu(D_f) = 0$, where D_f is the set of points where f is discontinuous.

Proof. (5) $\Longrightarrow$ (1): Immediate.

(5) $\Longrightarrow$ (2): This follows by setting $f = 1_A$, so that $D_f = \partial A$, and $\mu(D_f) = \mu(\partial A) = 0$. Then $\mu_n(A) = \int f \, d\mu_n \to \int f \, d\mu = \mu(A)$.

(2) $\Longrightarrow$ (3): Immediate, since the boundary of $(-\infty, x]$ is $\{x\}$.

(1) $\Longrightarrow$ (3): Let $\epsilon > 0$, and let f be the function defined by $f(t) = 1$ for $t \leq x$, $f(t) = 0$ for $t \geq x + \epsilon$, with f linear on the interval $(x, x + \epsilon)$ (see Figure 10.1.2 (a)). Then f is continuous, with $1_{(-\infty, x]} \leq f \leq 1_{(-\infty, x+\epsilon]}$. Hence,

$$\limsup_{n \to \infty} \mu_n((-\infty, x]) \leq \limsup_{n \to \infty} \int f d\mu_n = \int f d\mu \leq \mu((-\infty, x + \epsilon]).$$

---

[*]In fact, it corresponds to the weak* ("weak-star") topology from functional analysis, with $\mathcal{X}$ the set of all continuous functions on $\mathbf{R}$ vanishing at infinity (cf. Exercise 10.3.8), with norm defined by $\|f\| = \sup_{x \in \mathbf{R}} |f(x)|$, and with dual space $\mathcal{X}^*$ consisting of all finite signed Borel measures on $\mathbf{R}$. The Helly Selection Principle below then follows from Alaoglu's Theorem. See e.g. pages 161–2, 205, and 216 of Folland (1984).

This is true for any $\epsilon > 0$, so we conclude that $\limsup_{n\to\infty} \mu_n\left((-\infty, x]\right) \leq \mu\left((-\infty, x]\right)$.

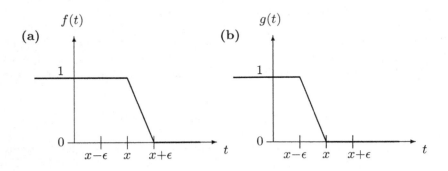

Figure 10.1.2. Functions used in proof of Theorem 10.1.1.

Similarly, if we let f be the function defined by $f(t) = 1$ for $t \leq x - \epsilon$, $f(t) = 0$ for $t \geq x$, with f linear on the interval $(x - \epsilon, x)$ (see Figure 10.1.2 (b)), then $1_{(-\infty, x-\epsilon]} \leq f \leq 1_{(-\infty, x]}$, and we obtain that

$$\liminf_{n\to\infty} \mu_n\left((-\infty, x]\right) \geq \liminf_{n\to\infty} \int f \, d\mu_n = \int f \, d\mu \geq \mu\left((-\infty, x - \epsilon]\right).$$

This is true for any $\epsilon > 0$, so we conclude that $\liminf_{n\to\infty} \mu_n\left((-\infty, x]\right) \geq \mu\left((-\infty, x)\right)$.

But if $\mu\{x\} = 0$, then $\mu\left((-\infty, x]\right) = \mu\left((-\infty, x)\right)$, so we must have

$$\limsup_{n\to\infty} \mu_n\left((-\infty, x]\right) = \liminf_{n\to\infty} \mu_n\left((-\infty, x]\right) = \mu\left((-\infty, x]\right),$$

as claimed.

(3) $\implies$ (4): We first define the cumulative distribution functions, by $F_n(x) = \mu_n\left((-\infty, x]\right)$ and $F(x) = \mu\left((-\infty, x]\right)$. Then, if we let $(\Omega, \mathcal{F}, \mathbf{P})$ be Lebesgue measure on $[0, 1]$, and let $Y_n(\omega) = \inf\{x; F_n(x) \geq \omega\}$ and $Y(\omega) = \inf\{x; F(x) \geq \omega\}$, then as in Lemma 7.1.2 we have $\mathcal{L}(Y_n) = \mu_n$ and $\mathcal{L}(Y) = \mu$. Note that if $F(z) < a$, then $Y(a) \geq z$, while if $F(z) \geq b$, then $Y(b) \leq z$.

Since $\{F_n\} \to F$ at most points, it seems reasonable that $\{Y_n\} \to Y$ at most points. We will prove that $\{Y_n\} \to Y$ at points of continuity of Y. Then, since Y is non-decreasing, it can have at most a countable number of discontinuities: indeed, it has at most $m\big(Y(n+1) - Y(n)\big) < \infty$ discontinuities of size $\geq 1/m$ within the interval $(n, n+1]$, then take countable union over m and n. Since countable sets have Lebesgue measure 0, this implies that $\{Y_n\} \to Y$ with probability 1, proving (4).

Suppose, then, that Y is continuous at ω, and let $y = Y(\omega)$. For any $\epsilon > 0$, we claim that $F(y-\epsilon) < \omega < F(y+\epsilon)$. Indeed, if we had $F(y-\epsilon) = \omega$, then setting $w = y - \epsilon$ and $b = \omega$ above, this would imply $Y(\omega) \leq y - \epsilon = Y(\omega) - \epsilon$, a contradiction. Or, if we had $F(y+\epsilon) = \omega$, then setting $z = y+\epsilon$ and $a = \omega + \delta$ above, this would imply $Y(\omega + \delta) \geq y + \epsilon = Y(\omega) + \epsilon$ for all $\delta > 0$, contradicting the continuity of Y at ω. So, $F(y - \epsilon) < \omega < F(y + \epsilon)$ for all $\epsilon > 0$.

Next, given $\epsilon > 0$, find ϵ' with $0 < \epsilon' < \epsilon$ such that $\mu\{y - \epsilon'\} = \mu\{y + \epsilon'\} = 0$. Then $F_n(y - \epsilon') \to F(y - \epsilon')$ and $F_n(y + \epsilon') \to F(y + \epsilon')$, so $F_n(y - \epsilon') < \omega < F_n(y + \epsilon')$ for all sufficiently large n. This in turn implies (setting first $z = y - \epsilon'$ and $a = \omega$ above, and then $w = y + \epsilon'$ and $b = \omega$ above) that $y - \epsilon' \leq Y_n(\omega) \leq y + \epsilon'$, i.e. $|Y_n(\omega) - Y(\omega)| \leq \epsilon' < \epsilon$ for all sufficiently large n. Hence, $Y_n(\omega) \to Y(\omega)$.

(4) $\implies$ (5): Recall that if f is continuous at x, and if $\{x_n\} \to x$, then $f(x_n) \to f(x)$. Hence, if $\{Y_n\} \to Y$ and $Y \notin D_f$, then $\{f(Y_n)\} \to f(Y)$. It follows that $\mathbf{P}[\{f(Y_n)\} \to f(Y)] \geq \mathbf{P}[\{Y_n\} \to Y \text{ and } Y \notin D_f]$. But by assumption, $\mathbf{P}[\{Y_n\} \to Y] = 1$ and $\mathbf{P}[Y \notin D_f] = \mu(D_f^C) = 1$, so also $\mathbf{P}[\{f(Y_n)\} \to f(Y)] = 1$. If f is bounded, then from the bounded convergence theorem, $\mathbf{E}[f(Y_n)] \to \mathbf{E}[f(Y)]$, i.e. $\int f \, d\mu_n \to \int f \, d\mu$, as claimed. ∎

For a first example, let μ be Lebesgue measure on $[0,1]$, and let μ_n be defined by $\mu_n\left(\frac{i}{n}\right) = \frac{1}{n}$ for $i = 1, 2, \ldots, n$. Then μ is purely continuous while μ_n is purely discrete; furthermore, $\mu(\mathbf{Q}) = 0$ while $\mu_n(\mathbf{Q}) = 1$ for each n. On the other hand, for any $0 \leq x \leq 1$, we have $\mu\left((-\infty, x]\right) = x$ while $\mu_n\left((-\infty, x]\right) = \lfloor nx \rfloor / n$. Hence, $|\mu_n\left((-\infty, x]\right) - \mu\left((-\infty, x]\right)| \leq \frac{1}{n} \to 0$ as $n \to \infty$, so we do indeed have $\mu_n \Rightarrow \mu$. (Note that $\partial \mathbf{Q} = [0,1]$ so that $\mu(\partial \mathbf{Q}) \neq 0$.)

For a second example, suppose $X_1, X_2, \ldots$ are i.i.d. with finite mean m, and $S_n = \frac{1}{n}(X_1 + \ldots + X_n)$. Then the weak law of large numbers says that for any $\epsilon > 0$ we have $\mathbf{P}(S_n \leq m - \epsilon) \to 0$ and $\mathbf{P}(S_n \leq m + \epsilon) \to 1$ as $n \to \infty$. It follows that $\mathcal{L}(S_n) \Rightarrow \delta_m(\cdot)$, a point mass at m. Note that it is *not* necessarily the case that $\mathbf{P}(S_n \leq m) \to \delta_m\left((-\infty, m]\right) = 1$, but this is no contradiction since the boundary of $(-\infty, m]$ is $\{m\}$, and $\delta_m\{m\} \neq 0$.

10.2. Connections to other convergence.

We now explore a sufficient condition for weak convergence.

Proposition 10.2.1. *If $\{X_n\} \to X$ in probability, then $\mathcal{L}(X_n) \Rightarrow \mathcal{L}(X)$.*

Proof. For any $\epsilon > 0$, if $X > z + \epsilon$ and $|X_n - X| < \epsilon$, then we must have $X_n > z$. That is, $\{X > z + \epsilon\} \cap \{|X_n - X| < \epsilon\} \subseteq \{X_n > z\}$. Taking complements, $\{X \leq z + \epsilon\} \cup \{|X_n - X| \geq \epsilon\} \supseteq \{X_n \leq z\}$. Hence, by the order-preserving property and subadditivity, $\mathbf{P}(X_n \leq z) \leq \mathbf{P}(X \leq z + \epsilon) + \mathbf{P}(|X - X_n| \geq \epsilon)$. Since $\{X_n\} \to X$ in probability, we get that $\limsup_{n\to\infty} \mathbf{P}(X_n \leq z) \leq \mathbf{P}(X \leq z + \epsilon)$. Letting $\epsilon \searrow 0$ gives $\limsup_{n\to\infty} \mathbf{P}(X_n \leq z) \leq \mathbf{P}(X \leq z)$.

Similarly, interchanging X and X_n and replacing z with $z - \epsilon$ in the above gives $\mathbf{P}(X \leq z - \epsilon) \leq \mathbf{P}(X_n \leq z) + \mathbf{P}(|X - X_n| \geq \epsilon)$, or $\mathbf{P}(X_n \leq z) \geq \mathbf{P}(X \leq z - \epsilon) - \mathbf{P}(|X - X_n| \geq \epsilon)$, so $\liminf \mathbf{P}(X_n \leq z) \geq \mathbf{P}(X \leq z - \epsilon)$. Letting $\epsilon \searrow 0$ gives $\liminf \mathbf{P}(X_n \leq z) \geq \mathbf{P}(X < z)$.

If $\mathbf{P}(X = z) = 0$, then $\mathbf{P}(X < z) = \mathbf{P}(X \leq z)$, so we must have $\liminf \mathbf{P}(X_n \leq z) = \limsup \mathbf{P}(X_n \leq z) = \mathbf{P}(X \leq z)$, as claimed. ∎

Remark. We sometimes write $\mathcal{L}(X_n) \Rightarrow \mathcal{L}(X)$ simply as $X_n \Rightarrow X$, and say that $\{X_n\}$ converges weakly (or, *in distribution*) to X.

We now have an interesting near-circle of implications. We already knew (Proposition 5.2.3) that if $X_n \to X$ almost surely, then $X_n \to X$ in probability. We now see from Proposition 10.2.1 that this in turn implies that $\mathcal{L}(X_n) \Rightarrow \mathcal{L}(X)$. And from Theorem 10.1.1(4), this implies that there are random variables Y_n and Y having the same laws, such that $Y_n \to Y$ almost surely.

Note that the converse to Proposition 10.2.1 is clearly false, since the fact that $\mathcal{L}(X_n) \Rightarrow \mathcal{L}(X)$ says nothing about the underlying relationship between X_n and X, it only says something about their laws. For example, if $X, X_1, X_2, \ldots$ are i.i.d., each equal to ± 1 with probability $\frac{1}{2}$, then of course $\mathcal{L}(X_n) \Rightarrow \mathcal{L}(X)$, but on the other hand $\mathbf{P}(|X - X_n| \geq 2) = \frac{1}{2} \not\to 0$, so X_n does *not* converge to X in probability or with probability 1. However, if X is *constant* then the converse to Proposition 10.2.1 does hold (Exercise 10.3.1).

Finally, we note that Skorohod's Theorem may be used to translate results involving convergence with probability 1 to results involving weak convergence (or, by Proposition 10.2.1, convergence in probability). For example, we have

Proposition 10.2.2. *Suppose* $\mathcal{L}(X_n) \Rightarrow \mathcal{L}(X)$, *with* $X_n \geq 0$. *Then* $\mathbf{E}(X) \leq \liminf \mathbf{E}(X_n)$.

Proof. By Skorohod's Theorem, we can find random variables Y_n and Y with $\mathcal{L}(Y_n) = \mathcal{L}(X_n)$, $\mathcal{L}(Y) = \mathcal{L}(X)$, and $Y_n \to Y$ with probability 1. Then, from Fatou's Lemma,

$$\mathbf{E}(X) = \mathbf{E}(Y) = \mathbf{E}\left(\liminf Y_n\right) \leq \liminf \mathbf{E}(Y_n) = \liminf \mathbf{E}(X_n).$$ ∎

For example, if $X \equiv 0$, and if $\mathbf{P}(X_n = n) = \frac{1}{n}$ and $\mathbf{P}(X_n = 0) = 1 - \frac{1}{n}$, then $\mathcal{L}(X_n) \Rightarrow \mathcal{L}(X)$, and $0 = \mathbf{E}(X) \leq \liminf \mathbf{E}(X_n) = 1$. (In fact, here $X_n \to X$ in probability, as well.)

Remark 10.2.3. We note that most of these weak convergence concepts have direct analogues for higher-dimensional distributions, not considered here; see e.g. Billingsley (1995, Section 29).

10.3. Exercises.

Exercise 10.3.1. Suppose $\mathcal{L}(X_n) \Rightarrow \delta_c$ for some $c \in \mathbf{R}$. Prove that $\{X_n\}$ converges to c in probability.

Exercise 10.3.2. Let $X, Y_1, Y_2, \ldots$ be independent random variables, with $\mathbf{P}(Y_n = 1) = 1/n$ and $\mathbf{P}(Y_n = 0) = 1 - 1/n$, . Let $Z_n = X + Y_n$. Prove that $\mathcal{L}(Z_n) \Rightarrow \mathcal{L}(X)$, i.e. that the law of Z_n converges weakly to the law of X.

Exercise 10.3.3. Let $\mu_n = N(0, \frac{1}{n})$ be a normal distribution with mean 0 and variance $\frac{1}{n}$. Does the sequence $\{\mu_n\}$ converge weakly to some probability measure? If yes, to what measure?

Exercise 10.3.4. Prove that weak limits, if they exist, are *unique*. That is, if $\mu, \nu, \mu_1, \mu_2, \ldots$ are probability measures, and $\mu_n \Rightarrow \mu$, and also $\mu_n \Rightarrow \nu$, then $\mu = \nu$.

Exercise 10.3.5. Let μ_n be the **Poisson**(n) distribution, and let μ be the **Poisson**(5) distribution. Show explicitly that each of the five conditions of Theorem 10.1.1 are violated.

Exercise 10.3.6. Let $a_1, a_2, \ldots$ be any sequence of non-negative real numbers with $\sum_i a_i = 1$. Define the discrete measure μ by $\mu(\cdot) = \sum_{i \in \mathbf{N}} a_i \delta_i(\cdot)$, where $\delta_i(\cdot)$ is a point-mass at the positive integer i. Construct a sequence $\{\mu_n\}$ of probability measures, each having a density with respect to Lebesgue measure, such that $\mu_n \Rightarrow \mu$.

Exercise 10.3.7. Let $\mathcal{L}(Y) = \mu$, where μ has continuous density f. For $n \in \mathbf{N}$, let $Y_n = \lfloor nY \rfloor / n$, and let $\mu_n = \mathcal{L}(Y_n)$.
(a) Describe μ_n explicitly.
(b) Prove that $\mu_n \Rightarrow \mu$.
(c) Is μ_n discrete, or absolutely continuous, or neither? What about μ?

Exercise 10.3.8.   Prove that the following are equivalent.
(1) $\mu_n \Rightarrow \mu$.
(2) $\int f \, d\mu_n \to \int f \, d\mu$ for all *non-negative* bounded continuous $f : \mathbf{R} \to \mathbf{R}$.
(3) $\int f \, d\mu_n \to \int f \, d\mu$ for all non-negative continuous $f : \mathbf{R} \to \mathbf{R}$ with *compact support*, i.e. such that there are finite a and b with $f(x) = 0$ for all $x < a$ and all $x > b$.
(4) $\int f \, d\mu_n \to \int f \, d\mu$ for all continuous $f : \mathbf{R} \to \mathbf{R}$ with compact support.
(5) $\int f \, d\mu_n \to \int f \, d\mu$ for all non-negative continuous $f : \mathbf{R} \to \mathbf{R}$ which *vanish at infinity*, i.e. such that $\lim_{x \to -\infty} f(x) = \lim_{x \to \infty} f(x) = 0$.
(6) $\int f \, d\mu_n \to \int f \, d\mu$ for all continuous $f : \mathbf{R} \to \mathbf{R}$ which vanish at infinity.
[Hints: You may assume the fact that all continuous functions on $\mathbf{R}$ which have compact support or vanish at infinity are bounded. Then, showing that (1) $\implies$ each of (4)–(6), and that each of (4)–(6) $\implies$ (3), is easy. For (2) $\implies$ (1), note that if $|f| \leq M$, then $f + M$ is non-negative. For (3) $\implies$ (2), note that if f is non-negative bounded continuous and $m \in \mathbf{Z}$, then $f_m \equiv f \, \mathbf{1}_{[m,m+1)}$ is non-negative bounded with compact support and is "nearly" continuous; then recall Figure 10.1.2, and that $f = \sum_{m \in \mathbf{Z}} f_m$.]

Exercise 10.3.9.   Let $0 < M < \infty$, and let $f, f_1, f_2, \ldots : [0, 1] \to [0, M]$ be Borel-measurable functions with $\int_0^1 f \, d\lambda = \int_0^1 f_n \, d\lambda = 1$. Suppose $\lim_n f_n(x) = f(x)$ for each fixed $x \in [0, 1]$. Define probability measures $\mu, \mu_1, \mu_2, \ldots$ by $\mu(A) = \int_A f \, d\lambda$ and $\mu_n(A) = \int_A f_n \, d\lambda$, for Borel $A \subseteq [0, 1]$. Prove that $\mu_n \Rightarrow \mu$.

Exercise 10.3.10.   Let $f : [0, 1] \to (0, \infty)$ be a continuous function such that $\int_0^1 f \, d\lambda = 1$ (where λ is Lebesgue measure on $[0, 1]$). Define probability measures μ and $\{\mu_n\}$ by $\mu(A) = \int_0^1 f \, \mathbf{1}_A \, d\lambda$ and $\mu_n(A) = \sum_{i=1}^n f(i/n) \, \mathbf{1}_A(i/n) \, / \, \sum_{i=1}^n f(i/n)$.
(a) Prove that $\mu_n \Rightarrow \mu$. [Hint: Recall Riemann sums from calculus.]
(b) Explicitly construct random variables Y and $\{Y_n\}$ so that $\mathcal{L}(Y) = \mu$, $\mathcal{L}(Y_n) = \mu_n$, and $Y_n \to Y$ with probability 1. [Hint: Remember the proof of Theorem 10.1.1.]

10.4. Section summary.

This section introduced the notion of weak convergence. It proved equivalences of weak convergence in terms of convergence of expectations of bounded continuous functions, convergence of probabilities, convergence of cumulative distribution functions, and the existence of corresponding random variables which converge with probability 1. It proved that if random variables converge in probability (or with probability 1), then their laws converge weakly.

Weak convergence will be very important in the next section, including allowing for a precise statement and proof of the Central Limit Theorem (Theorem 11.2.2 on page 134).

11. Characteristic functions.

Given a random variable X, we define its *characteristic function* (or *Fourier transform*) by

$$\phi_X(t) = \mathbf{E}(e^{itX}) = \mathbf{E}[\cos(tX)] + i\,\mathbf{E}[\sin(tX)], \qquad t \in \mathbf{R}.$$

The characteristic function is thus a function from the real numbers to the complex numbers. Of course, by the Change of Variable Theorem (Theorem 6.1.1), $\phi_X(t)$ depends only on the *distribution* of X. We sometimes write $\phi_X(t)$ as $\phi(t)$.

The characteristic function is clearly very similar to the moment generating function introduced earlier; the only difference is the appearance of the imaginary number $i = \sqrt{-1}$ in the exponent. However, this change is significant; since $|e^{itX}| = 1$ for any (real) t and X, the triangle inequality implies that $|\phi_X(t)| \leq 1 < \infty$ for all t and all random variables X. This is quite a contrast to the case for moment generating functions, which could be infinite for any $s \neq 0$.

Like for moment generating functions, we have $\phi_X(0) = 1$ for any X, and if X and Y are independent then $\phi_{X+Y}(t) = \phi_X(t)\,\phi_Y(t)$ by (4.2.7). We further note that, with $\mu = \mathcal{L}(X)$, we have

$$|\phi_X(t+h) - \phi_X(t)| = \left| \int \left(e^{i(t+h)x} - e^{itx} \right) \mu(dx) \right|$$

$$\leq \int \left| e^{i(t+h)x} - e^{itx} \right| \mu(dx) \leq \int \left| e^{itx} \right| \left| e^{ihx} - 1 \right| \mu(dx)$$

$$= \int \left| e^{ihx} - 1 \right| \mu(dx).$$

Now, as $h \to 0$, this last quantity decreases to 0 by the bounded convergence theorem (since $|e^{ihx} - 1| \leq 2$). We conclude that ϕ_X is always a (uniformly) continuous function.

The derivatives of ϕ_X are also straightforward. The following proposition is somewhat similar to the corresponding result for $M_X(s)$ (Theorem 9.3.3), except that here we do not require a severe condition like "$M_X(s) < \infty$ for all $|s| < s_0$".

Proposition 11.0.1. *Suppose X is a random variable with $\mathbf{E}\left(|X|^k\right) < \infty$. Then for $0 \leq j \leq k$, ϕ_X has finite jth derivative, given by $\phi_X^{(j)}(t) = \mathbf{E}\left[(iX)^j e^{itX}\right]$. In particular, $\phi_X^{(j)}(0) = i^j \mathbf{E}(X^j)$.*

Proof. We proceed by induction on j. The case $j = 0$ is trivial. Assume now that the statement is true for $j - 1$. For $t \in \mathbf{R}$, let $F_t = (iX)^{j-1} e^{itX}$, so that $|F'_t| = |(iX)^j e^{itX}| = |X|^j$. Since $E(|X|^k) < \infty$, therefore also $\mathbf{E}(|X|^j) < \infty$. It thus follows from Proposition 9.2.1 that

$$\phi_X^{(j)}(t) = \frac{d}{dt}\phi_X^{(j-1)}(t) = \frac{d}{dt}\mathbf{E}[(iX)^{j-1}e^{itX}]$$

$$= \mathbf{E}[\frac{\partial}{\partial t}(iX)^{j-1}e^{itX}] = \mathbf{E}[(iX)^j e^{itX}]. \qquad \blacksquare$$

11.1. The continuity theorem.

In this subsection we shall prove the continuity theorem for characteristic functions (Theorem 11.1.14), which says that if characteristic functions converge pointwise, then the corresponding distributions converge weakly: $\mu_n \Rightarrow \mu$ if and only if $\phi_n(t) \to \phi(t)$ for all t. This is a very important result; for example, it is used to prove the central limit theorem in the next subsection. Unfortunately, the proof is somewhat technical; we must show that characteristic functions completely determine the corresponding distribution (Theorem 11.1.1 and Corollary 11.1.7 below), and must also establish a simple criterion for weak convergence of "tight" measures (Corollary 11.1.11).

We begin with an inversion theorem, which tells how to recover information about a probability distribution from its characteristic function.

Theorem 11.1.1. *(Fourier inversion theorem) Let μ be a Borel probability measure on $\mathbf{R}$, with characteristic function $\phi(t) = \int_\mathbf{R} e^{itx}\mu(dx)$. Then if $a < b$ and $\mu\{a\} = \mu\{b\} = 0$, then*

$$\mu([a,b]) = \lim_{T\to\infty} \frac{1}{2\pi} \int_{-T}^{T} \frac{e^{-ita} - e^{-itb}}{it}\phi(t)dt.$$

To prove Theorem 11.1.1, we use two computational lemmas.

Lemma 11.1.2. *For $T \geq 0$ and $a < b$,*

$$\int_\mathbf{R} \int_{-T}^{T} \left| \frac{e^{-ita} - e^{-itb}}{it}e^{itx} \right| dt\,\mu(dx) \leq 2T(b-a) < \infty.$$

Proof. We first note by the triangle inequality that

$$\left| \frac{e^{-ita} - e^{-itb}}{it}e^{itx} \right| = \left| \frac{e^{-ita} - e^{-itb}}{it} \right| = \left| \int_a^b e^{-itr}dr \right|$$

$$\leq \int_a^b \left| e^{-itr} \right| dr = \int_a^b 1 \, dr = b - a \, .$$

Hence,

$$\int_{\mathbf{R}} \int_{-T}^T \left| \frac{e^{-ita} - e^{-itb}}{it} e^{itx} \right| dt \, \mu(dx) \leq \int_{\mathbf{R}} \int_{-T}^T (b - a) \, dt \, \mu(dx)$$

$$= \int_{\mathbf{R}} 2 \, T \, (b - a) \, \mu(dx) \; = \; 2 \, T \, (b - a) \, . \qquad \blacksquare$$

Lemma 11.1.3. *For $T \geq 0$ and $\theta \in \mathbf{R}$,*

$$\lim_{T \to \infty} \int_{-T}^T \frac{\sin(\theta t)}{t} dt \; = \; \pi \operatorname{sign}(\theta) \, , \qquad (11.1.4)$$

where $\operatorname{sign}(\theta) = 1$ *for* $\theta > 0$, $\operatorname{sign}(\theta) = -1$ *for* $\theta < 0$, *and* $\operatorname{sign}(0) = 0$. *Furthermore, there is* $M < \infty$ *such that* $|\int_{-T}^T [\sin(\theta t)/t] \, dt| \leq M$ *for all* $T \geq 0$ *and* $\theta \in \mathbf{R}$.

Proof (optional). When $\theta = 0$ both sides of (11.1.4) vanish, so assume $\theta \neq 0$. Making the substitution $s = |\theta| t$, $dt = ds/|\theta|$ gives

$$\int_{-T}^T \frac{\sin(\theta t)}{t} dt \; = \; \operatorname{sign}(\theta) \int_{-T}^T \frac{\sin(|\theta| t)}{t} dt \; = \; \operatorname{sign}(\theta) \int_{-|\theta|T}^{|\theta|T} \frac{\sin s}{s} ds \, ,$$
$$(11.1.5)$$

and hence

$$\lim_{T \to \infty} \int_{-T}^T \frac{\sin(\theta s)}{s} ds \; = \; 2 \operatorname{sign}(\theta) \int_0^\infty \frac{\sin s}{s} ds \, . \qquad (11.1.6)$$

Furthermore,

$$\int_0^\infty \frac{\sin s}{s} ds \; = \; \int_0^\infty (\sin s) \left(\int_0^\infty e^{-us} du \right) ds$$

$$= \int_0^\infty \left(\int_0^\infty (\sin s) e^{-us} ds \right) du \; = \; \int_0^\infty \left(\int_0^\infty (\sin s) e^{-us} ds \right) du \, .$$

Now, for $u > 0$, integrating by parts twice,

$$I_u \; \equiv \; \int_0^\infty (\sin s) e^{-us} ds \; = \; (-\cos s) e^{-us} \Big|_{s=0}^\infty - \int_0^\infty (-\cos s)(-u) e^{-us} ds$$

$$= \; 0 - (-1) - u \left((\sin s) e^{-us} \Big|_{s=0}^\infty + \int_0^\infty (\sin s)(-u) e^{-us} ds \right)$$

$$= 1 + 0 - 0 - u^2 \int_0^\infty (\sin s) e^{-us} ds.$$

Hence, $I_u = 1 - u^2 I_u$, so $I_u = 1/(1 + u^2)$.

We then compute that

$$\int_0^\infty \frac{\sin s}{s} ds = \int_0^\infty I_u \, du = \int_0^\infty \frac{1}{1 + u^2} \, du = \arctan(u) \Big|_{u=0}^\infty$$

$$= \arctan(\infty) - \arctan(0) = \pi/2 - 0 = \pi/2.$$

Combining this with (11.1.6) gives (11.1.4).

Finally, since convergent sequences are bounded, it follows from (11.1.4) that the set $\left\{ \int_{-T}^T [\sin(t)/t] \, dt \right\}_{T \geq 0}$ is bounded. It then follows from (11.1.5) that the set $\left\{ \int_{-T}^T [\sin(\theta t)/t] \, dt \right\}_{T \geq 0, \, \theta \in \mathbf{R}}$ is bounded as well. ∎

Proof of Theorem 11.1.1. We compute that

$$\frac{1}{2\pi} \int_{-T}^T \frac{e^{-ita} - e^{-itb}}{it} \phi(t) \, dt$$

$$= \frac{1}{2\pi} \int_{-T}^T \frac{e^{-ita} - e^{-itb}}{it} \left(\int_{\mathbf{R}} e^{itx} \mu(dx) \right) dt \qquad \text{[by definition of } \phi(t)]$$

$$= \frac{1}{2\pi} \int_{\mathbf{R}} \int_{-T}^T \frac{e^{it(x-a)} - e^{it(x-b)}}{it} \, dt \, \mu(dx) \qquad \text{[by Fubini and Lemma 11.1.2]}$$

$$= \frac{1}{2\pi} \int_{\mathbf{R}} \int_{-T}^T \frac{\sin(t(x-a)) - \sin(t(x-b))}{t} \, dt \, \mu(dx) \qquad \text{[since } \frac{\cos(ct)}{t} \text{ is odd]}.$$

Hence, we may use Lemma 11.1.3, together with the bounded convergence theorem and Remark 9.1.9, to conclude that

$$\lim_{T \to \infty} \frac{1}{2\pi} \int_{-T}^T \frac{e^{-ita} - e^{-itb}}{it} \phi(t) \, dt$$

$$= \frac{1}{2\pi} \int_{-\infty}^\infty \pi \left[\operatorname{sign}(x - a) - \operatorname{sign}(x - b) \right] \mu(dx)$$

$$= \int_{-\infty}^\infty \frac{1}{2} \left[\operatorname{sign}(x - a) - \operatorname{sign}(x - b) \right] \mu(dx)$$

$$= \frac{1}{2} \mu\{a\} + \mu((a, b)) + \frac{1}{2} \mu\{b\}.$$

(The last equality follows because $(1/2)[\operatorname{sign}(x - a) - \operatorname{sign}(x - b)]$ is equal to 0 if $x < a$ or $x > b$; is equal to $1/2$ if $x = a$ or $x = b$; and is equal to 1 if $a < x < b$.) But if $\mu\{a\} = \mu\{b\} = 0$, then this is precisely equal to $\mu([a, b])$, as claimed. ∎

From this theorem easily follows the important

Corollary 11.1.7. *(Fourier uniqueness theorem) Let X and Y be random variables. Then $\phi_X(t) = \phi_Y(t)$ for all $t \in \mathbf{R}$ if and only if $\mathcal{L}(X) = \mathcal{L}(Y)$, i.e. if and only if X and Y have the same distribution.*

Proof. Suppose $\phi_X(t) = \phi_Y(t)$ for all $t \in \mathbf{R}$. From the theorem, we know that $\mathbf{P}(a \leq X \leq b) = \mathbf{P}(a \leq Y \leq b)$ provided that $\mathbf{P}(X = a) = \mathbf{P}(X = b) = \mathbf{P}(Y = a) = \mathbf{P}(Y = b) = 0$, i.e. for all but countably many choices of a and b. But then by taking limits and using continuity of probabilities, we see that $\mathbf{P}(X \in I) = \mathbf{P}(Y \in I)$ for all intervals $I \subseteq \mathbf{R}$. It then follows from uniqueness of extensions (Proposition 2.5.8) that $\mathcal{L}(X) = \mathcal{L}(Y)$.

Conversely, if $\mathcal{L}(X) = \mathcal{L}(Y)$, then Corollary 6.1.3 implies that $\mathbf{E}(e^{itX}) = \mathbf{E}(e^{itY})$, i.e. $\phi_X(t) = \phi_Y(t)$ for all $t \in \mathbf{R}$. ∎

This last result makes the continuity theorem at least plausible. However, to prove the continuity theorem we require some further results.

Lemma 11.1.8. *(Helly Selection Principle) Let $\{F_n\}$ be a sequence of cumulative distribution functions (i.e. $F_n(x) = \mu_n((-\infty, x])$ for some probability distribution μ_n). Then there is a subsequence $\{F_{n_k}\}$, and a non-decreasing right-continuous function F with $0 \leq F \leq 1$, such that $\lim_{k \to \infty} F_{n_k}(x) = F(x)$ for all $x \in \mathbf{R}$ such that F is continuous at x. [On the other hand, we might not have $\lim_{x \to -\infty} F(x) = 0$ or $\lim_{x \to \infty} F(x) = 1$.]*

Proof. Since the rationals are countable, we can write them as $\mathbf{Q} = \{q_1, q_2, \ldots\}$. Since $0 \leq F_n(q_1) \leq 1$ for all n, the Bolzano-Weierstrass theorem (see page 204) says there is at least one subsequence $\{\ell_k^{(1)}\}$ such that $\lim_{k \to \infty} F_{\ell_k(1)}(q_1)$ exists. Then, there is a further subsequence $\{\ell_k^{(2)}\}$ (i.e., $\{\ell_k^{(2)}\}$ is a subsequence of $\{\ell_k^{(1)}\}$) such that $\lim_{k \to \infty} F_{\ell_k(2)}(q_2)$ exists (but also $\lim_{k \to \infty} F_{\ell_k(2)}(q_1)$ exists, since $\{\ell_k^{(2)}\}$ is a subsequence of $\{\ell_k^{(1)}\}$). Continuing, for each $m \in \mathbf{N}$ there is a further subsequence $\{\ell_k^{(m)}\}$ such that $\lim_{k \to \infty} F_{\ell_k(m)}(q_j)$ exists for $j \leq m$.

We now define the subsequence we want by $n_k = \ell_k^{(k)}$, i.e. we take the kth element of the kth subsequence. (This trick is called the *diagonalisation method.*) Since $\{n_k\}$ is a subsequence of $\{\ell_k\}$ from the kth point onwards, this ensures that $\lim_{k \to \infty} F_{n_k}(q) \equiv G(q)$ exists for each $q \in \mathbf{Q}$. Since each F_{n_k} is non-decreasing, therefore G is also non-decreasing.

To continue, we set $F(x) = \inf\{G(q); \ q \in \mathbf{Q}, \ q > x\}$. Then F is easily seen to be non-decreasing, with $0 \leq F(x) \leq 1$. Furthermore, F is right-continuous, since if $\{x_n\} \searrow x$ then $\{\{q \in \mathbf{Q} : q > x_n\}\} \nearrow \{q \in \mathbf{Q} : q > x\}$, and hence $F(x_n) \to F(x)$ as in Exercise 3.6.4. Also, since G is non-decreasing, we have $F(q) \geq G(q)$ for all $q \in \mathbf{Q}$.

Now, if F is continuous at x, then given $\epsilon > 0$ we can find rational numbers r, s, and u with $r < u < x < s$, and with $F(s) - F(r) < \epsilon$. We then note that

$$
\begin{aligned}
F(x) - \epsilon \; &\leq \; F(r) \\
&= \; \inf_{q>r} G(q) \\
&= \; \inf_{q>r} \lim_{k} F_{n_k}(q) \\
&= \; \inf_{q>r} \liminf_{k} F_{n_k}(q) \\
&\leq \; \liminf_{k} F_{n_k}(u) \qquad \text{since } u \in \mathbf{Q}, \; u > r \\
&\leq \; \liminf_{k} F_{n_k}(x) \qquad \text{since } x > u \\
&\leq \; \limsup_{k} F_{n_k}(x) \\
&\leq \; \limsup_{k} F_{n_k}(s) \qquad \text{since } s > x \\
&= \; G(s) \\
&\leq \; F(s) \\
&\leq \; F(x) + \epsilon .
\end{aligned}
$$

This is true for any $\epsilon > 0$, hence we must have

$$
\liminf_{k} F_{n_k}(x) = \limsup_{k} F_{n_k}(x) = F(x) \,,
$$

so that $\lim_k F_{n_k}(x) = F(x)$, as claimed. ∎

Unfortunately, Lemma 11.1.8 does not ensure that $\lim_{x \to \infty} F(x) = 1$ or $\lim_{x \to -\infty} F(x) = 0$ (see e.g. Exercise 11.5.1). To rectify this, we require a new notion. We say that a collection $\{\mu_n\}$ of probability measures on $\mathbf{R}$ is *tight* if for all $\epsilon > 0$, there are $a < b$ with $\mu_n([a, b]) \geq 1 - \epsilon$ for all n. That is, all of the measures give most of their mass to the same finite interval; mass does not "escape off to infinity".

Exercise 11.1.9. Prove that:
(a) any *finite* collection of probability measures is tight.
(b) the *union* of two tight collections of probability measures is tight.
(c) any sub-collection of a tight collection is tight.

We then have the following.

Theorem 11.1.10. *If $\{\mu_n\}$ is a tight sequence of probability measures, then there is a subsequence $\{\mu_{n_k}\}$ and a probability measure μ, such that $\mu_{n_k} \Rightarrow \mu$, i.e. $\{\mu_{n_k}\}$ converges weakly to μ.*

Proof. Let $F_n(x) = \mu_n\left((-\infty, x]\right)$. Then by Lemma 11.1.8, there is a subsequence F_{n_k} and a function F such that $F_{n_k}(x) \to F(x)$ at all continuity points of F. Furthermore $0 \le F \le 1$.

We now claim that F is actually a probability distribution function, i.e. that $\lim_{x\to-\infty} F(x) = 0$ and $\lim_{x\to\infty} F(x) = 1$. Indeed, let $\epsilon > 0$. Then using tightness, we can find points $a < b$ which are continuity points of F, such that $\mu_n\left((a, b]\right) \ge 1 - \epsilon$ for all n. But then

$$\lim_{x\to\infty} F(x) \;-\; \lim_{x\to-\infty} F(x) \ge F(b) - F(a)$$

$$= \lim_k \left[F_{n_k}(b) - F_{n_k}(a)\right] = \lim_k \mu_{n_k}\left((a, b]\right) \ge 1 - \epsilon\,.$$

This is true for all $\epsilon > 0$, so we must have $\lim_{x\to\infty} F(x) - \lim_{x\to-\infty} F(x) = 1$, proving the claim.

Hence, F is indeed a probability distribution function. Thus, we can define the probability measure μ by $\mu\left((a, b]\right) = F(b) - F(a)$ for $a < b$. Then $\mu_{n_k} \Rightarrow \mu$ by Theorem 10.1.1, and we are done. ∎

A main use of this theorem comes from the following corollary.

Corollary 11.1.11. *Let $\{\mu_n\}$ be a tight sequence of probability distributions on $\mathbf{R}$. Suppose that μ is the only possible weak limit of $\{\mu_n\}$, in the sense that whenever $\mu_{n_k} \Rightarrow \nu$ then $\nu = \mu$ (that is, whenever a subsequence of the $\{\mu_n\}$ converges weakly to some probability measure, then that probability measure must be μ). Then $\mu_n \Rightarrow \mu$, i.e. the full sequence converges weakly to μ.*

Proof. If $\mu_n \not\Rightarrow \mu$, then by Theorem 10.1.1, it is *not* the case that $\mu_n(\infty, x] \to \mu(-\infty, x]$ for all $x \in \mathbf{R}$ with $\mu\{x\} = 0$. Hence, we can find $x \in \mathbf{R}$, $\epsilon > 0$, and a subsequence $\{n_k\}$, with $\mu\{x\} = 0$, but with

$$\left|\mu_{n_k}\left((-\infty, x]\right) - \mu\left((-\infty, x]\right)\right| \ge \epsilon, \qquad k \in \mathbf{N}. \qquad (11.1.12)$$

On the other hand, $\{\mu_{n_k}\}$ is a subcollection of $\{\mu_n\}$ and hence tight, so by Theorem 11.1.10 there is a further subsequence $\{\mu_{n_{k_j}}\}$ which converges weakly to some probability measure, say ν. But then by hypothesis we must have $\nu = \mu$, which is a contradiction to (11.1.12). ∎

Corollary 11.1.11 is nearly the last thing we need to prove the continuity theorem. We require just one further result, concerning a sufficient condition for a sequence of measures to be tight.

Lemma 11.1.13. *Let $\{\mu_n\}$ be a sequence of probability measures on $\mathbf{R}$, with characteristic functions $\phi_n(t) = \int e^{itx} \mu_n(dx)$. Suppose there is a* *function g which is continuous at 0, such that $\lim_n \phi_n(t) = g(t)$ for each $|t| < t_0$ for some $t_0 > 0$. Then $\{\mu_n\}$ is tight.*

Proof.   We first note that $g(0) = \lim_n \phi_n(0) = \lim_n 1 = 1$. We then compute that, for $y > 0$,

$$
\begin{aligned}
\mu_n\left(\left(-\infty, -\tfrac{2}{y}\right] \cup \left[\tfrac{2}{y}, \infty\right)\right) &= \int_{|x| \geq 2/y} 1\, \mu_n(dx) \\
&\leq 2 \int_{|x| \geq 2/y} \left(1 - \tfrac{1}{y|x|}\right) \mu_n(dx) \\
&\leq 2 \int_{|x| \geq 2/y} \left(1 - \tfrac{\sin(yx)}{yx}\right) \mu_n(dx) \\
&= \int_{|x| \geq 2/y} (1/y) \int_{-y}^{y} \left(1 - \cos(tx)\right) dt\, \mu_n(dx) \\
&\leq \int_{x \in \mathbf{R}} (1/y) \int_{-y}^{y} \left(1 - \cos(tx)\right) dt\, \mu_n(dx) \\
&= \int_{x \in \mathbf{R}} (1/y) \int_{-y}^{y} (1 - e^{itx}) dt\, \mu_n(dx) \\
&= \tfrac{1}{y} \int_{-y}^{y} (1 - \phi_n(t))\, dt .
\end{aligned}
$$

Here the first inequality uses that $1 - \frac{1}{y|x|} \geq \frac{1}{2}$ whenever $|x| \geq 2/y$, the second inequality uses that $\frac{\sin(yx)}{yx} = \frac{\sin(y|x|)}{y|x|} \leq \frac{1}{y|x|}$, the second equality uses that $\int_{-y}^{y} \cos(tx)\, dt = \frac{2\sin(yx)}{x}$, the final inequality uses that $1 - \cos(tx) \geq 0$, the third equality uses that $\int_{-y}^{y} \sin(tx)\, dt = 0$, and the final equality uses Fubini's theorem (which is justified since the function is bounded and hence has finite double-integral).

  To finish the proof, let $\epsilon > 0$. Since $g(0) = 1$ and g is continuous at 0, we can find y_0 with $0 < y_0 < t_0$ such that $|1 - g(t)| \leq \epsilon/4$ whenever $|t| \leq y_0$. Then $\left|\frac{1}{y_0} \int_{-y_0}^{y_0} (1 - g(t)) dt\right| < \epsilon/2$. Now, $\phi_n(t) \to g(t)$ for all $|t| \leq y_0$, and $|\phi_n(t)| \leq 1$. Hence, by the bounded convergence theorem, we can find $n_0 \in \mathbf{N}$ such that $\left|\frac{1}{y_0} \int_{-y_0}^{y_0} (1 - \phi_n(t)) dt\right| < \epsilon$ for all $n \geq n_0$.

  Hence, $\mu_n\left(-\frac{2}{y_0}, \frac{2}{y_0}\right) = 1 - \mu_n\left((-\infty, -\frac{2}{y_0}] \cup [\frac{2}{y_0}, \infty)\right) > 1 - \epsilon$ for all $n \geq n_0$. It follows from the definition that $\{\mu_n\}$ is tight.   ∎

We are now, finally, in a position to prove the continuity theorem.

Theorem 11.1.14.   *(Continuity Theorem) Let $\mu, \mu_1, \mu_2, \ldots$ be probability measures, with corresponding characteristic functions $\phi, \phi_1, \phi_2, \ldots$. Then $\mu_n \Rightarrow \mu$ if and only if $\phi_n(t) \to \phi(t)$ for all $t \in \mathbf{R}$. In words, the probability measures $\{\mu_n\}$ converge weakly to μ if and only if their characteristic functions converge pointwise to that of μ.*

Proof.   First, suppose that $\mu_n \Rightarrow \mu$. Then, since $\cos(tx)$ and $\sin(tx)$ are bounded continuous functions, we have as $n \to \infty$ for each $t \in \mathbf{R}$ that

$$
\begin{aligned}
\phi_n(t) &= \int \cos(tx) \mu_n(dx) + i \int \sin(tx) \mu_n(dx) \\
&\to \int \cos(tx) \mu(dx) + i \int \sin(tx) \mu(dx) \\
&= \phi(t) .
\end{aligned}
$$

Conversely, suppose that $\phi_n(t) \to \phi(t)$ for each $t \in \mathbf{R}$. Then by Lemma 11.1.13 (with $g = \phi$), the $\{\mu_n\}$ are tight. Now, suppose that we have $\mu_{n_k} \Rightarrow \nu$ for some subsequence $\{\mu_{n_k}\}$ and some measure ν. Then, from the previous paragraph we must have $\phi_{n_k}(t) \to \phi_\nu(t)$ for all t, where $\phi_\nu(t) = \int e^{itx}\nu(dx)$. On the other hand, we know that $\phi_{n_k}(t) \to \phi(t)$ for all t; hence, we must have $\phi_\nu = \phi$. But from Fourier uniqueness (Corollary 11.1.7), this implies that $\nu = \mu$.

Hence, we have shown that μ is the only possible weak limit of the $\{\mu_n\}$. Therefore, from Corollary 11.1.11, we must have $\mu_n \Rightarrow \mu$, as claimed. ∎

11.2. The Central Limit Theorem.

Now that we have proved the continuity theorem (Theorem 11.1.14), it is very easy to prove the classical central limit theorem.

First, we compute the characteristic function for the standard normal distribution $N(0, 1)$, i.e. for a random variable X having density with respect to Lebesgue measure given by $f_X(x) = \frac{1}{\sqrt{2\pi}}e^{-x^2/2}$. That is, we wish to compute

$$\phi_X(t) = \int_{-\infty}^{\infty} e^{itx}\frac{1}{\sqrt{2\pi}}e^{-x^2/2}dx.$$

Comparing with the computation leading to (9.3.2), we might expect that $\phi_X(t) = M_X(it) = e^{(it)^2/2} = e^{-t^2/2}$. This is in fact correct, and can be justified using theory of complex analysis. But to avoid such technicalities, we instead resort to a trick.

Proposition 11.2.1. *If $X \sim N(0, 1)$, then $\phi_X(t) = e^{-t^2/2}$ for all $t \in \mathbf{R}$.*

Proof. By Proposition 9.2.1 (with $F_t = e^{itX}$ and $Y = |X|$, so that $\mathbf{E}(Y) < \infty$ and $|F_t'| = |(iX)e^{itX}| = |X| \le Y$ for all t), we can differentiate under the integral sign, to obtain that

$$\phi_X'(t) = \int_{-\infty}^{\infty} ix\, e^{itx}\frac{1}{\sqrt{2\pi}}e^{-x^2/2}dx = \int_{-\infty}^{\infty} i\, e^{itx}\frac{1}{\sqrt{2\pi}}x\, e^{-x^2/2}dx.$$

Integrating by parts gives that

$$\phi_X'(t) = \int_{-\infty}^{\infty} i(it)e^{itx}\frac{1}{\sqrt{2\pi}}e^{-x^2/2}dx = -t\,\phi_X(t).$$

Hence, $\phi_X'(t) = -t\,\phi_X(t)$, so that $\frac{d}{dt}\log\phi_X(t) = -t$. Also, we know that $\log\phi_X(0) = \log 1 = 0$. Hence, we must have $\log\phi_X(t) = \int_0^t (-s)\, ds = -t^2/2$, whence $\phi_X(t) = e^{-t^2/2}$.   ∎

We can now prove

Theorem 11.2.2.   *(Central Limit Theorem) Let $X_1, X_2, \ldots$ be i.i.d.*
with finite mean m and finite variance v. Set $S_n = X_1 + \ldots + X_n$. Then as
$n \to \infty$,

$$\mathcal{L}\left(\frac{S_n - nm}{\sqrt{vn}}\right) \Rightarrow \mu_N,$$

where $\mu_N = N(0,1)$ is the standard normal distribution, i.e. the distribution
having density $\frac{1}{\sqrt{2\pi}} e^{-x^2/2}$ with respect to Lebesgue measure.

Proof.   By replacing X_i by $\frac{X_i - m}{\sqrt{v}}$, we can (and do) assume that $m = 0$
and $v = 1$.
    Let $\phi_n(t) = \mathbf{E}\left(e^{itS_n/\sqrt{n}}\right)$ be the characteristic function of $S_n/\sqrt{n}$. By
the continuity theorem (Theorem 11.1.14), and by Proposition (11.2.1), it
suffices to show that $\lim_n \phi_n(t) = e^{-t^2/2}$ for each fixed $t \in \mathbf{R}$.
    To this end, set $\phi(t) = \mathbf{E}(e^{itX_1})$. Then as $n \to \infty$, using a Taylor
expansion and Proposition 11.0.1,

$$
\begin{aligned}
\phi_n(t) &= \mathbf{E}\left(e^{it(X_1 + \ldots + X_n)/\sqrt{n}}\right) \\
&= \phi(t/\sqrt{n})^n \\
&= \left(1 + \tfrac{it}{\sqrt{n}}\mathbf{E}(X_1) + \tfrac{1}{2!}\left(\tfrac{it}{\sqrt{n}}\right)^2 \mathbf{E}\left((X_1)^2\right) + o(1/n)\right)^n \\
&= \left(1 - \tfrac{t^2}{2n} + o(1/n)\right)^n \\
&\to e^{-t^2/2},
\end{aligned}
$$

as claimed. (Here $o(1/n)$ means a quantity q_n such that $q_n/(1/n) \to 0$ as
$n \to \infty$. Formally, the limit holds since for any $\epsilon > 0$, for sufficiently large
n we have $q_n \geq -\epsilon/n$ and also $q_n \leq \epsilon/n$, so that the lim inf is $\geq e^{-(t^2/2)-\epsilon}$
and the lim sup is $\leq e^{-(t^2/2)+\epsilon}$.)   ∎

    Since the normal distribution has no points of positive measure, this
theorem immediately implies (by Theorem 10.1.1) the simpler-seeming

Corollary 11.2.3.   *Let $\{X_n\}$, m, v, and S_n be as above. Then for each*
fixed $x \in R$,

$$\lim_{n \to \infty} \mathbf{P}\left(\frac{S_n - nm}{\sqrt{nv}} \leq x\right) = \Phi(x), \qquad (11.2.4)$$

where $\Phi(x) = \int_{-\infty}^{x} \frac{1}{\sqrt{2\pi}} e^{-t^2/2} dt$ is the cumulative distribution function for
the standard normal distribution.

This can also be written as $\mathbf{P}(S_n \leq nm + x\sqrt{nv}) \to \Phi(x)$. That is, $X_1 + \ldots + X_n$ is approximately equal to nm, with deviations from this value of order $\sqrt{n}$. For example, suppose $X_1, X_2, \ldots$ each have the **Poisson(5)** distribution. This implies that $m = \mathbf{E}(X_i) = 5$ and $v = \mathbf{Var}(X_i) = 5$. Hence, for each fixed $x \in \mathbf{R}$, we see that

$$\mathbf{P}\left(X_1 + \ldots + X_n \leq 5n + x\sqrt{5n}\right) \to \Phi(x), \qquad n \to \infty.$$

Remarks.

1. It is not essential in the Central Limit Theorem to divide by $\sqrt{v}$. Without doing so, the theorem asserts instead that

$$\mathcal{L}\left(\frac{S_n - nm}{\sqrt{n}}\right) \Rightarrow N(0, v).$$

2. Going backwards, the Central Limit Theorem in turn implies the WLLN, since if $y > m$, then as $n \to \infty$,

$$P[(S_n/n) \leq y] = P[S_n \leq ny] = P[(S_n - nm)/\sqrt{nv} \leq (ny - nm)/\sqrt{nv}]$$

$$\approx \Phi[(ny - nm)/\sqrt{nv}] = \Phi[\sqrt{n}(y - m)/\sqrt{v}] \to \Phi(+\infty) = 1,$$

and similarly if $y < m$ then $P[(S_n/n) \leq y] \to \Phi(-\infty) = 0$. Hence, $\mathcal{L}(S_n/n) \Rightarrow \delta_m$, and so S_n/n converges to m in probability.

11.3. Generalisations of the Central Limit Theorem.

The classical central limit theorem (Theorem 11.2.2 and Corollary 11.2.3) is extremely useful in many areas of science. However, it does have certain limitations. For example, it provides no quantitative bounds on the convergence in (11.2.4). Also, the insistence that the random variables be i.i.d. is sometimes too severe.

The first of these problems is solved by the Berry-Esseen Theorem, which states that if $X_1, X_2, \ldots$ are i.i.d. with finite mean m, finite positive variance v, and $\mathbf{E}\left(|X_i - m|^3\right) = \rho < \infty$, then

$$\left|\mathbf{P}\left(\frac{X_1 + \ldots + X_n - nm}{\sqrt{vn}} \leq x\right) - \Phi(x)\right| \leq \frac{3\rho}{\sqrt{nv^3}}.$$

This theorem thus provides a quantitative bound on the convergence in (11.2.4), depending only on the third moment. For a proof see e.g. Feller (1971, Section XVI.5). Note, however, that this error bound is absolute, not relative: as $x \to -\infty$, both $\mathbf{P}\left(\frac{X_1 + \ldots + X_n - nm}{\sqrt{vn}} \leq x\right)$ and $\Phi(x)$ get small, and the Berry-Esseen Theorem says less and less. In particular, the theorem does *not* assert that $\mathbf{P}\left(\frac{X_1+\ldots+X_n-nm}{\sqrt{vn}} \leq x\right)$ decreases as $O(e^{-x^2/2})$ as $x \to -\infty$, even though $\Phi(x)$ does.

Regarding the second problem, we mention just two of many results. To state them, we shall consider collections $\{Z_{nk}; n \geq 1, 1 \leq k \leq r_n\}$ of random variables such that each row $\{Z_{nk}\}_{1 \leq k \leq r_n}$ is independent, called *triangular arrays*. (If $r_n = n$ they form an actual triangle.) We shall assume for simplicity that $\mathbf{E}(Z_{nk}) = 0$ for each n and k. We shall further set $\sigma_{nk}^2 = \mathbf{E}(Z_{nk}^2)$ (assumed to be finite), $S_n = Z_{n1} + \ldots + Z_{nr_n}$, and $s_n^2 = \mathbf{Var}(S_n) = \sigma_{n1}^2 + \ldots + \sigma_{nr_n}^2$.

For such a triangular array, the Lindeberg Central Limit Theorem states that $\mathcal{L}(S_n/s_n) \Rightarrow N(0,1)$, provided that for each $\epsilon > 0$, we have

$$\lim_{n \to \infty} \frac{1}{s_n^2} \sum_{k=1}^{r_n} \mathbf{E}[Z_{nk}^2 \, \mathbf{1}_{|Z_{nk}| \geq \epsilon s_n}] = 0. \qquad (11.3.1)$$

This *Lindeberg condition* states, roughly, that as $n \to \infty$, the tails of the Z_{nk} contribute less and less to the variance of S_n.

Exercise 11.3.2.   Consider the special case where $r_n = n$, with $Z_{nk} = \frac{1}{\sqrt{nv}} Y_k$ where $\{Y_k\}$ are i.i.d. with mean 0 and variance $v < \infty$ (so $s_n = 1$).
(a) Prove that the Lindeberg condition (11.3.1) is satisfied in this case. [Hint: Use the Dominated Convergence Theorem.]
(b) Prove that the Lindeberg CLT implies Theorem 11.2.2.

This raises the question that, if (11.3.1) is *not* satisfied, then what other limiting distributions may arise? Call a distribution μ a *possible limit* if there exists a triangular array as defined above, with $\sup_n s_n^2 < \infty$ and $\lim_{n \to \infty} \max_{1 \leq k \leq r_n} \sigma_{nk}^2 = 0$ (so that no one term dominates the contribution to $\mathbf{Var}(S_n)$), such that $\mathcal{L}(S_n) \Rightarrow \mu$. Then we can ask, what distributions are possible limits? Obviously the normal distributions $N(0, v)$ are possible limits; indeed $\mathcal{L}(S_n) \Rightarrow N(0, v)$ whenever (11.3.1) is satisfied and $s_n^2 \to v$. But what else?

The answer is that the possible limits are precisely the infinitely divisible distributions having mean 0 and finite variance. Here a distribution μ is called *infinitely divisible* if for all $n \in \mathbf{N}$, there is a distribution ν_n such that the n-fold convolution of ν_n equals μ (in symbols: $\nu_n * \nu_n * \ldots * \nu_n = \mu$). Recall that this means that, if $X_1, X_2, \ldots, X_n \sim \nu_n$ are independent, then $X_1 + \ldots + X_n \sim \mu$.

Half of this theorem is obvious; indeed, if μ is infinitely divisible, then we can take $r_n = n$ and $\mathcal{L}(X_{nk}) = \nu_n$ in the triangular array, to get that

$\mathcal{L}(S_n) \Rightarrow \mu$. For a proof of the converse, see e.g. Billingsley (1995, Theorem 28.2).

11.4. Method of moments.

There is another way of proving weak convergence of probability measures, which does not explicitly use characteristic functions or the continuity theorem (though its proof of correctness does, through Corollary 11.1.11). Instead, it uses *moments*, as we now discuss.

Recall that a probability distribution μ on **R** has moments defined by $\alpha_k = \int x^k \mu(dx)$, for $k = 1, 2, 3, \ldots$. Suppose these moments all exist and are all finite. Then is μ the *only* distribution having precisely these moments? And, if a sequence $\{\mu_n\}$ of distributions have moments which *converge* to those of μ, then does it follow that $\mu_n \Rightarrow \mu$, i.e. that the μ_n converges weakly to μ? We shall see in this section that such conclusions hold sometimes, but not always.

We shall say that a distribution μ is *determined by its moments* if all its moments are finite, and if no other distribution has identical moments. (That is, we have $\int |x^k| \mu(dx) < \infty$ for all $k \in \mathbf{N}$, and furthermore whenever $\int x^k \mu(dx) = \int x^k \nu(dx)$ for all $k \in \mathbf{N}$, then we must have $\nu = \mu$.)

We first show that, for those distributions determined by their moments, convergence of moments implies weak convergence of distributions; this result thus reduces the second question above to the first question.

Theorem 11.4.1.   *Suppose that μ is determined by its moments. Let $\{\mu_n\}$ be a sequence of distributions, such that $\int x^k \mu_n(dx)$ is finite for all $n, k \in \mathbf{N}$, and such that $\lim_{n\to\infty} \int x^k \mu_n(dx) = \int x^k \mu(dx)$ for each $k \in \mathbf{N}$. Then $\mu_n \Rightarrow \mu$, i.e. the μ_n converge weakly to μ.*

Proof.   We first claim that $\{\mu_n\}$ is tight. Indeed, since the moments converge to finite quantities, we can find $K_k \in \mathbf{R}$ with $\int x^k \mu_n(dx) \le K_k$ for all $n \in \mathbf{N}$. But then, by Markov's inequality, letting $Y_n \sim \mu_n$, we have

$$
\begin{aligned}
\mu_n([-R, R]) &= \mathbf{P}(|Y_n| \le R) \\
&= 1 - \mathbf{P}(|Y_n| > R) \\
&= 1 - \mathbf{P}(Y_n^2 > R^2) \\
&\ge 1 - (\mathbf{E}[Y_n^2] / R^2) \\
&\ge 1 - (K_2 / R^2),
\end{aligned}
$$

which is $\ge 1 - \epsilon$ whenever $R \ge \sqrt{K_2/\epsilon}$, thus proving tightness.

We now claim that if any subsequence $\{\mu_{n_r}\}$ converges weakly to some distribution ν, then we must have $\nu = \mu$. The theorem will then follow from Corollary 11.1.11.

Indeed, suppose $\mu_{n_r} \Rightarrow \nu$. By Skorohod's theorem, we can find random variables Y and $\{Y_r\}$ with $\mathcal{L}(Y) = \nu$ and $\mathcal{L}(Y_r) = \mu_{n_r}$, such that $Y_r \to Y$ with probability 1. But then also $Y_r^k \to Y^k$ with probability 1. Furthermore, for $k \in \mathbf{N}$ and $\alpha > 0$, we have

$$\mathbf{E}\left(|Y_r|^k \mathbf{1}_{|Y_r|^k \geq \alpha}\right) \leq \mathbf{E}\left(\frac{|Y_r|^{2k}}{\alpha} \mathbf{1}_{|Y_r|^k \geq \alpha}\right) \leq \frac{1}{\alpha}\mathbf{E}\left(|Y_r|^{2k}\right)$$

$$= \frac{1}{\alpha}\mathbf{E}\left((Y_r)^{2k}\right) \leq \frac{K_{2k}}{\alpha},$$

which is independent of r, and goes to 0 as $\alpha \to \infty$. Hence, the $\{Y_r^k\}$ are uniformly integrable. Thus, by Theorem 9.1.6, $\mathbf{E}(Y_r^k) \to \mathbf{E}(Y^k)$, i.e. $\int x^k \mu_{n_r}(dx) \to \int x^k \nu(dx)$.

But we already know that $\int x^k \mu_{n_r}(dx) \to \int x^k \mu(dx)$. Hence, the moments of ν and μ must coincide. And, since μ is determined by its moments, we must have $\nu = \mu$, as claimed. ∎

This theorem leads to the question of which distributions are determined by their moments. Unfortunately, not all distributions are, as the following exercise shows.

Exercise 11.4.2. Let $f(x) = \frac{1}{x\sqrt{2\pi}}e^{-(\log x)^2/2}$ for $x > 0$ (with $f(x) = 0$ for $x \leq 0$) be the density function for the random variable e^X, where $X \sim N(0,1)$. Let $g(x) = f(x)\left(1 + \sin(2\pi \log x)\right)$. Show that $g(x) \geq 0$ and that $\int x^k g(x)dx = \int x^k f(x)dx$ for $k = 0, 1, 2, \ldots$. [Hint: Consider $\int x^k f(x) \sin(2\pi \log x)dx$, and make the substitution $x = e^s e^k$, $dx = e^s e^k ds$.] Show further that $\int |x|^k f(x)dx < \infty$ for all $k \in \mathbf{N}$. Conclude that g is a probability density function, and that g gives rise to the same (finite) moments as does f. Relate this to Theorem 11.4.1 above and Theorem 11.4.3 below.

On the other hand, if a distribution satisfies that its moment generating function is finite in a neighbourhood of the origin, then it will be determined by its moments, as we now show. (Unfortunately, the proof requires a result from complex function theory.)

Theorem 11.4.3. Let $s_0 > 0$, and let X be a random variable with moment generating function $M_X(s)$ which is finite for $|s| < s_0$. Then $\mathcal{L}(X)$ is determined by its moments (and also by $M_X(s)$).

Proof (optional). Let $f_X(z) = \mathbf{E}(e^{zX})$ for $z \in \mathbf{C}$. Since $|e^{zX}| = e^{X \, \Re e \, z}$, we see that $f_X(z)$ is finite whenever $|\Re e \, z| < s_0$. Furthermore, just like for $M_X(s)$, it follows that $f_X(z)$ is *analytic* on $\{z \in \mathbf{C};\ |\Re e \, z| < s_0\}$. Now, if

Y has the same moments as does X, then for $|s| < s_0$, we have by order-preserving and countable linearity that

$$\mathbf{E}\left(e^{sY}\right) \leq \mathbf{E}\left(e^{sY} + e^{-sY}\right)$$

$$= 2\,\mathbf{E}\left(1 + \frac{s^2 Y^2}{2!} + \dots\right) = 2\left(1 + \frac{s^2 \mathbf{E}(Y^2)}{2!} + \dots\right)$$

$$= 2\left(1 + \frac{s^2 \mathbf{E}(X^2)}{2!} + \dots\right) = M_X(s) + M_X(-s) < \infty.$$

Hence, $M_Y(s) < \infty$ for $|s| < s_0$. It now follows from Theorem 9.3.3 that $M_Y(s) = M_X(s)$ for $|s| < s_0$, i.e. that $f_X(s) = f_Y(s)$ for real $|s| < s_0$. By the uniqueness of analytic continuation, this implies that $f_Y(z) = f_X(z)$ for $|\Re e\, z| < s_0$. In particular, since $\phi_X(t) = f_X(it)$ and $\phi_Y(t) = f_Y(it)$, we have $\phi_X = \phi_Y$. Hence, by the uniqueness theorem for characteristic functions (Theorem 11.1.7), we must have $\mathcal{L}(Y) = \mathcal{L}(X)$, as claimed. ∎

Remark 11.4.4.   Proving weak convergence by showing convergence of moments is called the *method of moments*[*]. Indeed, it is possible to prove the central limit theorem in this manner, under appropriate assumptions.

Remark 11.4.5.   By similar reasoning, it is possible to show that if $M_{X_n}(s) < \infty$ and $M_X(s) < \infty$ for all $n \in \mathbf{N}$ and $|s| < s_0$, and also $M_{X_n}(s) \to M_X(s)$ for all $|s| < s_0$, then we must have $\mathcal{L}(X_n) \Rightarrow \mathcal{L}(X)$.

11.5. Exercises.

Exercise 11.5.1.   Let $\mu_n = \delta_n$ be a point mass at n (for $n = 1, 2, \dots$).
(a) Is $\{\mu_n\}$ tight?
(b) Does there exist a subsequence $\{\mu_{n_k}\}$, and a Borel probability measure μ, such that $\mu_{n_k} \Rightarrow \mu$? (If so, then specify $\{n_k\}$ and μ.) Relate this to theorems from this section.
(c) Setting $F_n(x) = \mu_n\left((-\infty, x]\right)$, does there exist a non-decreasing, right-continuous function F such that $F_n(x) \to F(x)$ for all continuity points x of F? (If so, then specify F.) Relate this to the Helly Selection Principle.
(d) Repeat part (c) for the case where $\mu_n = \delta_{-n}$ is a point mass at $-n$.

Exercise 11.5.2.   Let $\mu_n = \delta_{n \bmod 3}$ be a point mass at $n \bmod 3$. (Thus, $\mu_1 = \delta_1$, $\mu_2 = \delta_2$, $\mu_3 = \delta_0$, $\mu_4 = \delta_1$, $\mu_5 = \delta_2$, $\mu_6 = \delta_0$, etc.)

---

[*]This should not be confused with the statistical estimation procedure of the same name, which estimates unknown parameters by choosing them to make observed moments equal theoretical ones.

(a) Is $\{\mu_n\}$ tight?

(b) Does there exist a Borel probability measure μ, such that $\mu_n \Rightarrow \mu$? (If so, then specify μ.)

(c) Does there exist a subsequence $\{\mu_{n_k}\}$, and a Borel probability measure μ, such that $\mu_{n_k} \Rightarrow \mu$? (If so, then specify $\{n_k\}$ and μ.)

(d) Relate parts (b) and (c) to theorems from this section.

Exercise 11.5.3. Let $\{x_n\}$ be any sequence of points in the interval $[0, 1]$. Let $\mu_n = \delta_{x_n}$ be a point mass at x_n.

(a) Is $\{\mu_n\}$ tight?

(b) Does there exist a subsequence $\{\mu_{n_k}\}$, and a Borel probability measure μ, such that $\mu_{n_k} \Rightarrow \mu$? (Hint: by compactness, there must be a subsequence of points $\{x_{n_k}\}$ which converges, say to $y \in [0, 1]$. Then what does μ_{n_k} converge to?)

Exercise 11.5.4. Let $\mu_{2n} = \delta_0$, and let $\mu_{2n+1} = \delta_n$, for $n = 0, 1, 2, \ldots$.

(a) Does there exist a Borel probability measure μ, such that $\mu_n \Rightarrow \mu$?

(b) Suppose for some subsequence $\{\mu_{n_k}\}$ and some Borel probability measure ν, we have $\mu_{n_k} \Rightarrow \nu$. What must ν be?

(c) Relate parts (a) and (b) to Corollary 11.1.11. Why is there no contradiction?

Exercise 11.5.5. Let $\mu_n = \mathbf{Uniform}[0, n]$, so $\mu_n([a, b]) = (b - a)/n$ for $0 \leq a \leq b \leq n$.

(a) Prove or disprove that $\{\mu_n\}$ is tight.

(b) Prove or disprove that there is some probabilty measure μ such that $\mu_n \Rightarrow \mu$.

Exercise 11.5.6. Suppose $\mu_n \Rightarrow \mu$. Prove or disprove that $\{\mu_n\}$ must be tight.

Exercise 11.5.7. Define the Borel probability measure μ_n by $\mu_n(\{x\}) = 1/n$, for $x = 0, \frac{1}{n}, \frac{2}{n}, \ldots, \frac{n-1}{n}$. Let λ be Lebesgue measure on $[0, 1]$.

(a) Compute $\phi_n(t) = \int e^{itx} \mu_n(dx)$, the characteristic function of μ_n.

(b) Compute $\phi(t) = \int e^{itx} \lambda(dx)$, the characteristic function of λ.

(c) Does $\phi_n(t) \to \phi(t)$, for each $t \in \mathbf{R}$?

(d) What does the result in part (c) imply?

Exercise 11.5.8. Use characteristic functions to provide an alternative solution of Exercise 10.3.2.

Exercise 11.5.9. Use characteristic functions to provide an alternative solution of Exercise 10.3.3.

Exercise 11.5.10. Use characteristic functions to provide an alternative solution of Exercise 10.3.4.

Exercise 11.5.11. Compute the characteristic function $\phi_X(t)$, and also $\phi'_X(0) = i\mathbf{E}(X)$, where X follows
(a) the binomial distribution: $P(X = k) = \binom{n}{k}p^k(1-p)^{n-k}$, for $k = 0, 1, 2, \ldots, n$.
(b) the Poisson distribution: $P(X = k) = e^{-\lambda}\frac{\lambda^k}{k!}$, for $k = 0, 1, 2, \ldots$.
(c) the exponential distribution, with density with respect to Lebesgue measure given by $f_X(x) = \lambda e^{-\lambda x}$ for $x > 0$, and $f_X(x) = 0$ for $x < 0$.

Exercise 11.5.12. Suppose that for $n \in \mathbf{N}$, we have $\mathbf{P}[X_n = 5] = 1/n$ and $\mathbf{P}[X_n = 6] = 1 - (1/n)$.
(a) Compute the characteristic function $\phi_{X_n}(t)$, for all $n \in \mathbf{N}$ and $t \in \mathbf{R}$.
(b) Compute $\lim_{n\to\infty}\phi_{X_n}(t)$.
(c) Specify a distribution μ such that $\lim_{n\to\infty}\phi_{X_n}(t) = \int e^{itx}\mu(dx)$ for all $t \in \mathbf{R}$.
(d) Determine (with explanation) whether or not $\mathcal{L}(X_n) \Rightarrow \mu$.

Exercise 11.5.13. Let $\{X_n\}$ be i.i.d., each having mean 3 and variance 4. Let $S = X_1 + X_2 + \ldots + X_{10,000}$. In terms of $\Phi(x)$, give an approximate value for $\mathbf{P}[S \le 30,500]$.

Exercise 11.5.14. Let $X_1, X_2, \ldots$ be i.i.d. with mean 4 and variance 9. Find values $C(n, x)$, for $n \in \mathbf{N}$ and $x \in \mathbf{R}$, such that as $n \to \infty$, $\mathbf{P}[X_1 + X_2 + \ldots + X_n \le C(n, x)] \approx \Phi(x)$.

Exercise 11.5.15. Prove that the **Poisson**(λ) distribution, and the $N(m, v)$ (normal) distribution, are both infinitely divisible (for any $\lambda > 0$, $m \in \mathbf{R}$, and $v > 0$). [Hint: Use Exercises 9.5.15 and 9.5.16.]

Exercise 11.5.16. Let X be a random variable whose distribution $\mathcal{L}(X)$ is infinitely divisible. Let $a > 0$ and $b \in \mathbf{R}$, and set $Y = aX + b$. Prove that $\mathcal{L}(Y)$ is infinitely divisible.

Exercise 11.5.17. Prove that the **Poisson**(λ) distribution, the $N(m, v)$ distribution, and the **Exp**(λ) (exponential) distribution, are all determined by their moments, for any $\lambda > 0$, $m \in \mathbf{R}$, and $v > 0$.

Exercise 11.5.18. Let $X, X_1, X_2, \ldots$ be random variables which are uniformly bounded, i.e. there is $M \in \mathbf{R}$ with $|X| \le M$ and $|X_n| \le M$ for all n. Prove that $\{\mathcal{L}(X_n)\} \Rightarrow \mathcal{L}(X)$ if and only if $\mathbf{E}(X_n^k) \to \mathbf{E}(X^k)$ for all $k \in \mathbf{N}$.

11.6. Section summary.

This section introduced the characteristic function $\phi_X(t) = \mathbf{E}(e^{itX})$. After introducing its basic properties, it proved an Inversion Theorem (to recover the distribution of a random variable from its characteristic function) and a Uniqueness Theorem (which shows that if two random variables have the same characteristic function then they have the same distribution).

Then, using the Helly Selection Principle and the notion of tightness, it proved the important Continuity Theorem, which asserts the equivalence of weak convergence of distributions and pointwise convergence of characteristic functions. This important theorem was used to prove the Central Limit Theorem about weak convergence of averages of i.i.d. random variables to a normal distribution. Some generalisations of the Central Limit Theorem were briefly discussed.

The section ended with a discussion of the method of moments, an alternative way of proving weak convergence of random variables using only their moments, but not their characteristic functions.

12. Decomposition of probability laws.

Let μ be a Borel probability measure on $\mathbf{R}$. Recall that μ is *discrete* if it takes all its mass at individual points, i.e. if $\sum_{x \in \mathbf{R}} \mu\{x\} = \mu(\mathbf{R})$. Also μ is *absolutely continuous* if there is a non-negative Borel-measurable function f such that $\mu(A) = \int_A f(x)\lambda(dx) = \int \mathbf{1}_A(x)f(x)\lambda(dx)$ for all Borel sets A, where λ is Lebesgue measure on $\mathbf{R}$.

One could ask, are these the *only* possible types of probability distributions? Of course the answer to this question is no, as μ could be a *mixture* of a discrete and an absolutely continuous distribution, e.g. $\mu = \frac{1}{2}\delta_0 + \frac{1}{2}N(0,1)$. But can every probability distribution at least be written as such a mixture? Equivalently, is it true that every distribution μ with no discrete component (i.e. which satisfies $\mu\{x\} = 0$ for each $x \in \mathbf{R}$) must necessarily be absolutely continuous?

To examine this question, say that μ is *dominated* by λ (written $\mu \ll \lambda$), if $\mu(A) = 0$ whenever $\lambda(A) = 0$. Then clearly any absolutely continuous measure μ must be dominated by λ, since whenever $\lambda(A) = 0$ we would then have $\mu(A) = \int \mathbf{1}_A(x)f(x)\lambda(dx) = 0$. (We shall see in Corollary 12.1.2 below that in fact the converse to this statement also holds.)

On the other hand, suppose $Z_1, Z_2, \ldots$ are i.i.d. taking the value 1 with probability 2/3, and the value 0 with probability 1/3. Set $Y = \sum_{n=1}^{\infty} Z_n 2^{-n}$ (i.e. the base-2 expansion of Y is $0.Z_1 Z_2 \ldots$). Further, define $S \subseteq \mathbf{R}$ by

$$S = \left\{ x \in [0,1]; \ \lim_{n \to \infty} \frac{1}{n} \sum_{i=1}^{n} d_i(x) = \frac{2}{3} \right\},$$

where $d_i(x)$ is the ith digit in the (non-terminating) base-2 expansion of x. Then by the strong law of large numbers, we have $\mathbf{P}(Y \in S) = 1$ while $\lambda(S) = 0$. Hence, from the previous paragraph, the law of Y cannot be absolutely continuous. But clearly $\mathcal{L}(Y)$ has no discrete component. We conclude that $\mathcal{L}(Y)$ *cannot* be written as a mixture of a discrete and an absolutely continuous distribution. In fact, $\mathcal{L}(Y)$ is *singular* with respect to λ (written $\mathcal{L}(Y) \perp \lambda$), meaning that there is a subset $S \subseteq \mathbf{R}$ with $\lambda(S) = 0$ and $\mathbf{P}(Y \in S^C) = 0$.

12.1. Lebesgue and Hahn decompositions.

The main result of this section is the following.

Theorem 12.1.1. *(Lebesgue Decomposition) Any probability measure μ on $\mathbf{R}$ can uniquely be decomposed as $\mu = \mu_{disc} + \mu_{ac} + \mu_{sing}$, where the measures μ_{disc}, μ_{ac}, and μ_{sing} satisfy*

(a) μ_{disc} is discrete, i.e. $\sum_{x \in \mathbf{R}} \mu_{disc}\{x\} = \mu_{disc}(\mathbf{R})$;

(b) μ_{ac} is absolutely continuous, i.e. $\mu_{ac}(A) = \int_A f \, d\lambda$ for all Borel sets A, for some non-negative, Borel-measurable function f, where λ is Lebesgue measure on $\mathbf{R}$;

(c) μ_{sing} is singular continuous, i.e. $\mu_{sing}\{x\} = 0$ for all $x \in \mathbf{R}$, but there is $S \subseteq \mathbf{R}$ with $\lambda(S) = 0$ and $\mu_{sing}(S^C) = 0$.

Theorem 12.1.1 will be proved below. We first present an important corollary.

Corollary 12.1.2.   (Radon-Nikodym Theorem) A Borel probability measure μ is absolutely continuous (i.e. there is f with $\mu(A) = \int_A f \, d\lambda$ for all Borel A) if and only if it is dominated by λ (i.e. $\mu \ll \lambda$, i.e. $\mu(A) = 0$ whenever $\lambda(A) = 0$).

Proof.   We have already seen that if μ is absolutely continuous then $\mu \ll \lambda$.

For the converse, suppose $\mu \ll \lambda$, and let $\mu = \mu_{disc} + \mu_{ac} + \mu_{sing}$ as in Theorem 12.1.1. Since $\lambda\{x\} = 0$ for each x, we must have $\mu\{x\} = 0$ as well, so that $\mu_{disc} = 0$. Similarly, if S is such that $\lambda(S) = 0$ and $\mu_{sing}(S^C) = 0$, then we must have $\mu(S) = 0$, so that $\mu_{sing}(S) = 0$, so that $\mu_{sing} = 0$. Hence, $\mu = \mu_{ac}$, i.e. μ is absolutely continuous.   ∎

Remark 12.1.3.   If $\mu(A) = \int_A f \, d\lambda$ for all Borel A, we write $\frac{d\mu}{d\lambda} = f$, and call f the *density*, or *Radon-Nikodym derivative*, of μ with respect to λ.

It remains to prove Theorem 12.1.1. We begin with a lemma.

Lemma 12.1.4.   (Hahn Decomposition) Let ϕ be a finite "signed measure" on $(\Omega, \mathcal{F})$, i.e. $\phi = \mu - \nu$ for some finite measures μ and ν. Then there is a partition $\Omega = A^+ \,\dot{\cup}\, A^-$, with $A^+, A^- \in \mathcal{F}$, such that $\phi(E) \geq 0$ for all $E \subseteq A^+$, and $\phi(E) \leq 0$ for all $E \subseteq A^-$.

Proof.   Following Billingsley (1995, Theorem 32.1), we set

$$\alpha = \sup\{\phi(A); \, A \in \mathcal{F}\}.$$

We shall construct a subset A^+ such that $\phi(A^+) = \alpha$. Once we have done this, then we can then take $A^- = \Omega \setminus A^+$. Then if $E \subseteq A^+$ but $\phi(E) < 0$, then $\phi(A^+ \setminus E) = \phi(A^+) - \phi(E) > \phi(A^+) = \alpha$, which contradicts the definition of α. Similarly, if $E \subseteq A^-$ but $\phi(E) > 0$, then $\phi(A^+ \cup E) \geq \phi(A^+) + \phi(E) > \alpha$, again contradicting the definition of α. Thus, to complete the proof, it suffices to construct A^+ with $\phi(A^+) = \alpha$.

To that end, choose (by the definition of α) subsets $A_1, A_2, \ldots \in \mathcal{F}$ such that $\phi(A_n) \to \alpha$. Let $A = \bigcup A_i$, and let

$$\mathcal{G}_n = \left\{ \bigcap_{k=1}^n A_k', \text{ each } A_k' = A_k \text{ or } A \setminus A_k \right\}$$

(so that $\mathcal{G}_n$ contains $\leq 2^n$ different subsets, which are all disjoint). Then let

$$C_n = \bigcup_{\substack{S \in \mathcal{G}_n \\ \phi(S) \geq 0}} S,$$

i.e. C_n is the union of those elements of $\mathcal{G}_n$ with non-negative measure under ϕ. Finally, set $A^+ = \limsup C_n$. We claim that $\phi(A^+) = \alpha$.

First, note that since A_n is a union of certain particular elements of $\mathcal{G}_n$ (namely, all those formed with $A_n' = A_n$), and C_n is a union of all the ϕ-positive elements of $\mathcal{G}_n$, it follows that $\phi(C_n) \geq \phi(A_n)$.

Next, note that $C_m \cup \ldots \cup C_n$ is formed from $C_m \cup \ldots \cup C_{n-1}$ by including some additional ϕ-positive elements of $\mathcal{G}_n$, so $\phi(C_m \cup C_{m+1} \cup \ldots \cup C_n) \geq \phi(C_m \cup C_{m+1} \cup \ldots \cup C_{n-1})$. It follows by induction that $\phi(C_m \cup \ldots \cup C_n) \geq \phi(C_m) \geq \phi(A_m)$. Since this holds for all n, we must have (by continuity of probabilities on μ and ν separately) that $\phi(C_m \cup C_{m+1} \cup \ldots) \geq \phi(A_m)$ for all m.

But then (again by continuity of probabilities on μ and ν separately) we have

$$\phi(A^+) = \phi(\limsup C_n) = \lim_{m \to \infty} \phi(C_m \cup C_{m+1} \cup \ldots) \geq \lim_{m \to \infty} \phi(A_m) = \alpha.$$

Hence, $\phi(A^+) = \alpha$, as claimed. ∎

Remarks.

1. The Hahn decomposition is unique up to sets of ϕ-measure 0, i.e. if $A^+ \cup A^-$ and $B^+ \cup B^-$ are two Hahn decompositions then $\phi(A^+ \setminus B^+) = \phi(B^+ \setminus A^+) = \phi(A^- \setminus B^-) = \phi(B^- \setminus A^-) = 0$, since e.g. $A^+ \setminus B^+ = A^+ \cap B^-$ must have ϕ-measure both ≥ 0 and ≤ 0.

2. *Using* the Radon-Nikodym theorem, it is very easy to prove the Hahn decomposition; indeed, we can let $\xi = \mu + \nu$, let $f = \frac{d\mu}{d\xi}$ and $g = \frac{d\nu}{d\xi}$, and set $A^+ = \{f \geq g\}$. However, this reasoning would be circular, since we are going to use the Hahn decomposition to prove the Lebesgue decomposition which in turn proves the Radon-Nikodym theorem!

3. If ϕ is any countably additive mapping from $\mathcal{F}$ to $\mathbf{R}$ (not necessarily non-negative, nor assumed to be of the form $\mu - \nu$), then continuity of probabilities follows just as in Proposition 3.3.1, and the proof of Theorem 12.1.4 then goes through without change. Furthermore, it then

follows that $\phi = \mu - \nu$ where $\mu(E) = \phi(E \cap A^+)$ and $\nu(E) = -\phi(E \cap A^-)$, i.e. every such function is in fact a signed measure.

We are now able to prove our main theorem.

Proof of Theorem 12.1.1.   We first take care of μ_{disc}. Indeed, clearly we shall define $\mu_{disc}(A) = \sum_{x \in A} \mu\{x\}$, and then $\mu - \mu_{disc}$ has no discrete component. Hence, we assume from now on that μ has no discrete component.

Second, we note that by countable additivity, it suffices to assume that μ is supported on $[0, 1]$, so that we may also take λ to be Lebesgue measure on $[0, 1]$.

To continue, call a function g a *candidate density* if $g \geq 0$ and $\int_E g \, d\lambda \leq \mu(E)$ for all Borel sets E. We note that if g_1 and g_2 are candidate densities, then so is $\max(g_1, g_2)$, since

$$\int_E \max(g_1, g_2) \, d\lambda = \int_{E \cap \{g_1 \geq g_2\}} g_1 \, d\lambda + \int_{E \cap \{g_1 < g_2\}} g_2 \, d\lambda$$

$$\leq \mu(E \cap \{g_1 \geq g_2\}) + \mu(E \cap \{g_1 < g_2\}) = \mu(E).$$

Also, by the monotone convergence theorem, if $h_1, h_2, \ldots$ are candidate densities and $h_n \nearrow h$, then h is also a candidate density. It follows from these two observations that if $g_1, g_2, \ldots$ are candidate densities, then so is $\sup_n g_n = \lim_{n \to \infty} \max(g_1, \ldots, g_n)$.

Now, let $\beta = \sup\{\int_{[0,1]} g \, d\lambda; \ g$ a candidate density$\}$. Choose candidate densities g_n with $\int_{[0,1]} g_n d\lambda \geq \beta - \frac{1}{n}$, and let $f = \sup_{n \geq 1} g_n$, to obtain that f is a candidate density with $\int_{[0,1]} f \, d\lambda = \beta$, i.e. f is (up to a set of measure 0) the largest possible candidate density.

This f shall be our density for μ_{ac}. That is, we define $\mu_{ac}(A) = \int_A f \, d\lambda$. We then (of course) define $\mu_{sing}(A) = \mu(A) - \mu_{ac}(A)$. Since f was a candidate density, therefore $\mu_{sing}(A) \geq 0$. To complete the existence proof, it suffices to show that μ_{sing} is singular.

For each $n \in \mathbf{N}$, let $[0, 1] = A_n^+ \dot\cup A_n^-$ be a Hahn decomposition (cf. Lemma 12.1.4) for the signed measure $\phi_n = \mu_{sing} - \frac{1}{n}\lambda$. Set $M = \bigcup_n A_n^+$. Then $M^C = \bigcap_n A_n^-$, so that $M^C \subseteq A_n^-$ for each n. It follows that $(\mu_{sing} - \frac{1}{n}\lambda)(M^C) \leq 0$ for all n, so that $\mu_{sing}(M^C) \leq \frac{1}{n}\lambda(M^C)$ for all n. Hence, $\mu_{sing}(M^C) = 0$. We claim that $\lambda(M) = 0$, so that μ_{sing} is indeed singular. To prove this, we assume that $\lambda(M) > 0$, and derive a contradiction.

If $\lambda(M) > 0$, then there is $n \in \mathbf{N}$ with $\lambda(A_n^+) > 0$. For this n, we have $(\mu_{sing} - \frac{1}{n}\lambda)(E) \geq 0$, i.e. $\mu_{sing}(E) \geq \frac{1}{n}\lambda(E)$, for all $E \subseteq A_n^+$. We now claim that the function $g = f + \frac{1}{n}\mathbf{1}_{A_n^+}$ is a candidate density. Indeed, we compute for any Borel set E that

$$
\begin{aligned}
\int_E g\, d\lambda &= \int_E f\, d\lambda + \tfrac{1}{n}\int_E \mathbf{1}_{A_n^+}\, d\lambda \\
&= \mu_{ac}(E) + \tfrac{1}{n}\lambda(A_n^+ \cap E) \\
&\le \mu_{ac}(E) + \mu_{sing}(A_n^+ \cap E) \\
&\le \mu_{ac}(E) + \mu_{sing}(E) \\
&= \mu(E)\,,
\end{aligned}
$$

thus verifying the claim.

On the other hand, we have

$$
\int_{[0,1]} g\, d\lambda = \int_{[0,1]} f\, d\lambda + \frac{1}{n}\int_{[0,1]} \mathbf{1}_{A_n^+}\, d\lambda = \beta + \frac{1}{n}\lambda(A_n^+) > \beta\,,
$$

which contradicts the maximality of f. Hence, we must actually have had $\lambda(M) = 0$, showing that μ_{sing} must actually be singular, thus proving the existence part of the theorem.

Finally, we prove the uniqueness. Indeed, suppose $\mu = \mu_{ac} + \mu_{sing} = \nu_{ac} + \nu_{sing}$, with $\mu_{ac}(A) = \int_A f\, d\lambda$ and $\nu_{ac}(A) = \int_A g\, d\lambda$. Since μ_{sing} and ν_{sing} are singular, we can find S_1 and S_2 with $\lambda(S_1) = \lambda(S_2) = 0$ and $\mu_{sing}(S_1^C) = \nu_{sing}(S_2^C) = 0$. Let $S = S_1 \cup S_2$, and let $B = \{\omega \in S^C; f(\omega) < g(\omega)\}$. Then $g - f > 0$ on B, but $\int_B (g - f)d\lambda = \mu_{ac}(B) - \nu_{ac}(B) = \mu(B) - \nu(B) = 0$. Hence, $\lambda(B) = 0$. But we also have $\lambda(S) = 0$, hence $\lambda\{f < g\} = 0$. Similarly $\lambda\{f > g\} = 0$. We conclude that $\lambda\{f = g\} = 1$, whence $\mu_{ac} = \nu_{ac}$, whence $\mu_{sing} = \nu_{sing}$. ∎

Remark. Note that, while μ_{ac} is unique, the density $f = \frac{d\mu_{ac}}{d\lambda}$ is only unique up to a set of measure 0.

12.2. Decomposition with general measures.

Finally, we note that similar decompositions may be made with respect to other measures. Instead of considering absolute continuity or singularity with respect to λ, we can consider them with respect to any other probability measure ν, and the same proofs apply. Furthermore, by countable additivity, the above proofs go through virtually unchanged for σ-finite ν as well (cf. Remark 4.4.3). We state the more general results as follows. (For the general Radon-Nikodym theorem, we write $\mu \ll \nu$, and say that μ is *dominated* by ν, if $\mu(A) = 0$ whenever $\nu(A) = 0$.)

Theorem 12.2.1. *(Lebesgue Decomposition, general case)* If μ and ν are two σ-finite measures on some measurable space $(\Omega, \mathcal{F})$, then μ can uniquely be decomposed as $\mu = \mu_{ac} + \mu_{sing}$, where μ_{ac} is absolutely continuous with respect to ν (i.e., there is a non-negative measurable function $f : \Omega \to \mathbf{R}$ such that $\mu_{ac}(A) = \int_A f \, d\nu$ for all $A \in \mathcal{F}$), and μ_{sing} is singular (i.e., there is $S \subseteq \Omega$ with $\nu(S) = 0$ and $\mu_{sing}(S^C) = 0$).

Remark. In the above statement, for simplicity we avoid mentioning the discrete part μ_{disc}, thus allowing for the possibility that ν may itself have some discrete component (in which case the corresponding discrete component of μ would still be absolutely continuous with respect to ν). However, if ν has no discrete component, then if we wish we can extract μ_{disc} from μ_{sing} as in Theorem 12.1.1.

Corollary 12.2.2. (Radon-Nikodym Theorem, general case) If μ and ν are two σ-finite measures on some measurable space $(\Omega, \mathcal{F})$, then μ is absolutely continuous with respect to ν if and only if $\mu \ll \nu$.

If $\mu \ll \nu$, so $\mu(A) = \int_A f \, d\nu$ for all $A \in \mathcal{F}$, then we write $\frac{d\mu}{d\nu} = f$, and call $\frac{d\mu}{d\nu}$ the *Radon-Nikodym derivative* of μ with respect to ν. Thus, $\mu(A) = \int_A (\frac{d\mu}{d\nu}) \, d\nu$. If ν is a probability measure, then we can also write this as $\mu(A) = \mathbf{E}_\nu[\frac{d\mu}{d\nu} \mathbf{1}_A]$, where $\mathbf{E}_\nu$ stands for expectation with respect to ν.

12.3. Exercises.

Exercise 12.3.1. Prove that μ is discrete if and only if there is a countable set S with $\mu(S^C) = 0$.

Exercise 12.3.2. Let X and Y be discrete random variables (not necessarily independent), and let $Z = X + Y$. Prove that $\mathcal{L}(Z)$ is discrete.

Exercise 12.3.3. Let X be a random variable, and let $Y = cX$ for some constant $c > 0$.
(a) Prove or disprove that if $\mathcal{L}(X)$ is discrete, then $\mathcal{L}(Y)$ must be discrete.
(b) Prove or disprove that if $\mathcal{L}(X)$ is absolutely continuous, then $\mathcal{L}(Y)$ must be absolutely continuous.
(c) Prove or disprove that if $\mathcal{L}(X)$ is singular continuous, then $\mathcal{L}(Y)$ must be singular continuous.

Exercise 12.3.4. Let X and Y be random variables, with $\mathcal{L}(Y)$ absolutely continuous, and let $Z = X + Y$.
(a) Assume X and Y are independent. Prove that $\mathcal{L}(Z)$ is absolutely continuous, regardless of the nature of $\mathcal{L}(X)$. [Hint: Recall the convolution formula of Subsection 9.4.]

(b) Show that if X and Y are not independent, then $\mathcal{L}(Z)$ may fail to be absolutely continuous.

Exercise 12.3.5. Let X and Y be random variables, with $\mathcal{L}(X)$ discrete and $\mathcal{L}(Y)$ singular continuous. Let $Z = X + Y$. Prove that $\mathcal{L}(Z)$ is singular continuous. [Hint: If $\lambda(S) = \mathbf{P}(Y \in S^C) = 0$, consider the set $U = \{s + x : s \in S, \ \mathbf{P}(X = x) > 0\}$.]

Exercise 12.3.6. Let $A, B, Z_1, Z_2, \ldots$ be i.i.d., each equal to $+1$ with probability $2/3$, or equal to 0 with probability $1/3$. Let $Y = \sum_{i=1}^{\infty} Z_i \, 2^{-i}$ as at the beginning of this section (so $\nu \equiv L(Y)$ is singular continuous), and let $W \sim N(0,1)$. Finally, let $X = A(BY + (1 - B)W)$, and set $\mu = \mathcal{L}(X)$. Find a discrete measure μ_{disc}, an absolutely continuous measure μ_{ac}, and a singular continuous measure μ_s, such that $\mu = \mu_{disc} + \mu_{ac} + \mu_s$.

Exercise 12.3.7. Let μ, ν, and ρ be probability measures with $\mu \ll \nu \ll \rho$. Prove that $\frac{d\mu}{d\rho} = \frac{d\mu}{d\nu} \frac{d\nu}{d\rho}$ with ρ-probability 1. [Hint: Use Proposition 6.2.3.]

Exercise 12.3.8. Let μ and ν be probability measures with $\mu \ll \nu$ and $\nu \ll \mu$. (This is sometimes written as $\mu \equiv \nu$.) Prove that $\frac{d\mu}{d\nu} > 0$ with μ-probability 1, and in fact $\frac{d\nu}{d\mu} = 1 / \frac{d\mu}{d\nu}$.

Exercise 12.3.9. Let μ and ν be discrete probability measures, with $\phi = \mu - \nu$. Write down an *explicit* Hahn decomposition $\Omega = A^+ \,\dot{\cup}\, A^-$ for ϕ.

Exercise 12.3.10. Let μ and ν be absolutely continuous probability measures on $\mathbf{R}$ (with the Borel σ-algebra), with $\phi = \mu - \nu$. Write down an *explicit* Hahn decomposition $\mathbf{R} = A^+ \,\dot{\cup}\, A^-$ for ϕ.

12.4. Section summary.

This section proved the Lebesgue Decomposition theorem, which says that every measure may be uniquely written as the sum of a discrete measure, an absolutely continuous measure, and a singular continuous measure. From this followed the Radon-Nikodym Theorem, which gives a simple condition under which a measure is absolutely continuous.

13. Conditional probability and expectation.

Conditioning is a very important concept in probability, and we consider it here.

Of course, conditioning on events of *positive* measure is quite straightforward. We have already noted that if A and B are events, with $\mathbf{P}(B) > 0$, then we can define the conditional probability $\mathbf{P}(A \mid B) = \mathbf{P}(A \cap B) / \mathbf{P}(B)$; intuitively, this represents the probabilistic proportion of the event B which also includes the event A. More generally, if Y is a random variable, and if we define ν by $\nu(S) = \mathbf{P}(Y \in S \mid B) = \mathbf{P}(Y \in S, B) / \mathbf{P}(B)$, then $\nu = \mathcal{L}(Y \mid B)$ is a probability measure, called the *conditional distribution* of Y given B. We can then define conditional expectation by $\mathbf{E}(Y \mid B) = \int y \, \nu(dy)$. Also, $\mathcal{L}(Y \mathbf{1}_B) = P(B) \mathcal{L}(Y \mid B) + P(B^C) \delta_0$, so taking expectations and re-arranging,

$$\mathbf{E}(Y \mid B) = \mathbf{E}(Y \mathbf{1}_B) / \mathbf{P}(B). \tag{13.0.1}$$

No serious difficulties arise.

On the other hand, if $\mathbf{P}(B) = 0$ then this approach does not work at all. Indeed, it is quite unclear how to define something like $\mathbf{P}(Y \in S \mid B)$ in that case. Unfortunately, it frequently arises that we wish to condition on events of probability 0.

13.1. Conditioning on a random variable.

We begin with an example.

Example 13.1.1.   Let (X, Y) be uniformly distributed on the triangle $T = \{(x, y) \in \mathbf{R}^2; \ 0 \le y \le 2, \ y \le x \le 2\}$; see Figure 13.1.2. (That is, $\mathbf{P}((X, Y) \in S) = \frac{1}{2} \lambda_2(S \cap T)$ for Borel $S \subseteq \mathbf{R}^2$, where λ_2 is Lebesgue measure on $\mathbf{R}^2$; briefly, $d\mathbf{P} = \frac{1}{2} \mathbf{1}_T \, dx \, dy$.) Then what is $\mathbf{P}\left(Y > \frac{3}{4} \mid X = 1\right)$? What is $\mathbf{E}(Y \mid X = 1)$? Since $\mathbf{P}(X = 1) = 0$, it is not clear how to proceed. We shall return to this example below.

Because of this problem, we take a different approach. Given a random variable X, we shall consider conditional probabilities like $\mathbf{P}(A \mid X)$, and also conditional expected values like $\mathbf{E}(Y \mid X)$, to themselves be *random variables*. We shall think of them as functions of the "random" value X. This is very counter-intuitive: we are used to thinking of $\mathbf{P}(\cdots)$ and $\mathbf{E}(\cdots)$ as numbers, not random variables. However, we shall think of them as random variables, and we shall see that this allows us to partially resolve the difficulty of conditioning on sets of measure 0 (such as $\{X = 1\}$ above).

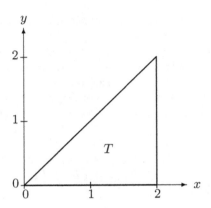

Figure 13.1.2. The triangle T in Example 13.1.1.

The idea is that, once we define these quantities to be random variables, then we can demand that they satisfy certain properties. For starters, we require that

$$\mathbf{E}\left[\mathbf{P}(A\,|\,X)\right] = \mathbf{P}(A)\,, \qquad \mathbf{E}\left[\mathbf{E}(Y\,|\,X)\right] = \mathbf{E}(Y)\,. \qquad (13.1.3)$$

In words, these random variables must have the correct *expected values*.

Unfortunately, this does not completely specify the distributions of the random variables $\mathbf{P}(A\,|\,X)$ and $\mathbf{E}(Y\,|\,X)$; indeed, there are infinitely many different distributions having the same mean. We shall therefore impose a stronger requirement. To state it, recall that if $\mathcal{G}$ is a sub-σ-algebra (i.e. a σ-algebra contained in the main σ-algebra $\mathcal{F}$), then a random variable Z is $\mathcal{G}$-measurable if $\{Z \le z\} \in \mathcal{G}$ for all $z \in \mathbf{R}$. (It follows that also $\{Z = z\} = \{Z \le z\} \setminus \bigcup_n \{Z \le z - \frac{1}{n}\} \in \mathcal{G}$.) Also, $\sigma(X) = \{\{X \in B\} : B \subseteq \mathbf{R}\ \text{Borel}\}$.

Definition 13.1.4. Given random variables X and Y with $\mathbf{E}|Y| < \infty$, and an event A, $\mathbf{P}(A\,|\,X)$ is a *conditional probability* of A given X if it is a $\sigma(X)$-measurable random variable and, for any Borel $S \subseteq \mathbf{R}$, we have

$$\mathbf{E}\left(\mathbf{P}(A\,|\,X)\mathbf{1}_{X \in S}\right) = \mathbf{P}\left(A \cap \{X \in S\}\right)\,. \qquad (13.1.5)$$

Similarly, $\mathbf{E}(Y\,|\,X)$ is a *conditional expectation* of Y given X if it is a $\sigma(X)$-

measurable random variable and, for any Borel $S \subseteq \mathbf{R}$, we have

$$\mathbf{E}\left(\mathbf{E}(Y \,|\, X) 1_{X \in S}\right) = \mathbf{E}\left(Y \, 1_{X \in S}\right). \qquad (13.1.6)$$

That is, we define conditional probabilities and expectations by specifying that certain expected values of them should assume certain values. This prompts some observations.

Remarks.

1. Requiring these quantities to be $\sigma(X)$-measurable means that they are functions of X alone, and do not otherwise depend on the sample point ω. Indeed, if a random variable Z is $\sigma(X)$-measurable, then for each $z \in \mathbf{R}$, $\{Z = z\} = \{X \in B_z\}$ for some Borel $B_z \subseteq \mathbf{R}$. Then $Z = f(X)$ where f is defined by $f(x) = z$ for all $x \in B_z$, i.e. Z is a function of X.

2. Of course, if we set $S = \mathbf{R}$ in (13.1.5) and (13.1.6), we obtain the special case (13.1.3).

3. Since expected values are unaffected by changes on a set of measure 0, we see that conditional probabilities and expectations are only unique up to a set of measure 0. Thus, if $\mathbf{P}(X = x) = 0$ for some particular value of x, then we may change $\mathbf{P}(A \,|\, X)$ on the set $\{X = x\}$ without restriction. However, we may only change its value on a set of measure 0; thus, for "most" values of X, the value $\mathbf{P}(A \,|\, X)$ cannot change. In this sense, we have mostly (but not entirely) overcome the difficulty of Example 13.1.1.

These conditional probabilities and expectations always exist:

Proposition 13.1.7. *Let X and Y be jointly defined random variables, and let A be an event. Then $\mathbf{P}(A \,|\, X)$ and $\mathbf{E}(Y \,|\, X)$ exist (though they are only unique up to a set of probability 0).*

Proof. We may define $\mathbf{P}(A \,|\, X)$ to be $\frac{d\nu}{d\mathbf{P}_0}$, where $\mathbf{P}_0$ and ν are measures on $\sigma(X)$ defined as follows. $\mathbf{P}_0$ is simply $\mathbf{P}$ restricted to $\sigma(X)$, and

$$\nu(E) = \mathbf{P}(A \cap E), \qquad E \in \sigma(X).$$

Note that $\nu \ll \mathbf{P}_0$, so by the general Radon-Nikodym Theorem (Corollary 12.2.2), $\frac{d\nu}{d\mathbf{P}_0}$ exists and is unique up to a set of probability 0.

Similarly, we may define $\mathbf{E}(Y \,|\, X)$ to be $\frac{d\rho^+}{d\mathbf{P}_0} - \frac{d\rho^-}{d\mathbf{P}_0}$, where

$$\rho^+(E) = \mathbf{E}\left(Y^+ 1_E\right), \qquad \rho^-(E) = \mathbf{E}\left(Y^- 1_E\right), \qquad E \in \sigma(X).$$

Then (13.1.5) and (13.1.6) are automatically satisfied, since e.g.

$$\mathbf{E}[\mathbf{P}(A \,|\, X) 1_{X \in S}] = \mathbf{E}\left[\frac{d\nu}{d\mathbf{P}_0} 1_{X \in S}\right] = \int_{X \in S} \frac{d\nu}{d\mathbf{P}_0} \, d\mathbf{P}$$

$$= \int_{X \in S} \frac{d\nu}{d\mathbf{P}_0} \, d\mathbf{P}_0 \; = \; \nu(\{X \in S\}) \; = \; \mathbf{P}(A \cap \{X \in S\}). \quad \blacksquare$$

Example 13.1.1 continued.   To better understand Definition 13.1.4, we return to the case where (X, Y) is assumed to be uniformly distributed over the triangle $T = \{(x, y) \in \mathbf{R}^2; \; 0 \le y \le 2, \; y \le x \le 2\}$. We can then define $\mathbf{P}(Y \ge \frac{3}{4} \,|\, X)$ and $\mathbf{E}(Y \,|\, X)$ to be what they "ought" to be, namely

$$\mathbf{P}\Big(Y \ge \frac{3}{4} \,\Big|\, X\Big) \; = \; \begin{cases} \frac{X - \frac{3}{4}}{X}, & X \ge 3/4 \\ 0, & X < 3/4 \end{cases} \; ; \qquad \mathbf{E}(Y \,|\, X) \; = \; \frac{X}{2}.$$

We then compute that, for example,

$$\mathbf{E}\Big(\mathbf{P}(Y \ge \frac{3}{4} \,|\, X) 1_{X \in S}\Big) \; = \; \int_{S \cap [\frac{3}{4}, 2]} \int_0^x \frac{x - \frac{3}{4}}{x} \frac{1}{2} \lambda(dy) \, \lambda(dx)$$

$$= \; \int_{S \cap [\frac{3}{4}, 2]} \Big(x - \frac{3}{4}\Big) \frac{1}{2} \lambda(dx)$$

while

$$\mathbf{P}\Big(Y \ge \frac{3}{4}, \; X \in S\Big) \; = \; \int_{S \cap [\frac{3}{4}, 2]} \int_{3/4}^x (1) \frac{1}{2} \lambda(dy) \, \lambda(dx)$$

$$= \; \int_{S \cap [\frac{3}{4}, 2]} \Big(x - \frac{3}{4}\Big) \frac{1}{2} \lambda(dx) \,,$$

thus verifying (13.1.5) for the case $A = \{Y \ge \frac{3}{4}\}$. Similarly, for $S \subseteq [0, 2]$,

$$\mathbf{E}\left(\mathbf{E}(Y \,|\, X) 1_{X \in S}\right) \; = \; \int_{T \cap (S \times \mathbf{R})} \frac{x}{2} \frac{1}{2} \lambda(dy) \, \lambda(dx)$$

$$= \; \int_S \frac{x}{2} \int_0^x \frac{1}{2} \lambda(dy) \, \lambda(dx) \; = \; \int_S \frac{x}{2} \frac{x}{2} \lambda(dx) \,,$$

while

$$\mathbf{E}\left(Y 1_{X \in S}\right) = \int_{T \cap (S \times \mathbf{R})} y \frac{1}{2} \lambda(dy) \, dx$$

$$= \; \int_S \int_0^x y \frac{1}{2} \lambda(dy) \, \lambda(dx) \; = \; \int_S \frac{x^2}{4} \lambda(dx) \,,$$

so that the two expressions in (13.1.6) are also equal.

   This example was quite specific, but similar ideas can be used more generally; see Exercises 13.4.3 and 13.4.4. In summary, conditional probabilities and expectations always exist, and are unique up to sets of probability 0, but finding them may require guessing appropriate random variables and then verifying that they satisfy (13.1.5) and (13.1.6).

13.2. Conditioning on a sub-σ-algebra.

So far we have considered conditioning on just a single random variable X. We may think of this, equivalently, as conditioning on the generated σ-algebra, $\sigma(X)$. More generally, we may wish to condition on something other than just a single random variable X. For example, we may wish to condition on a collection of random variables $X_1, X_2, \ldots, X_n$, or we may wish to condition on certain other events.

Given *any* sub-σ-algebra $\mathcal{G}$ (in place of $\sigma(X)$ above), we define the conditional probability $\mathbf{P}(A \mid \mathcal{G})$ and the conditional expectation $\mathbf{E}(Y \mid \mathcal{G})$ to be $\mathcal{G}$-measurable random variables satisfying that

$$\mathbf{E}\left(\mathbf{P}(A \mid \mathcal{G}) \mathbf{1}_G\right) = \mathbf{P}(A \cap G) \tag{13.2.1}$$

and

$$\mathbf{E}\left(\mathbf{E}(Y \mid \mathcal{G}) \mathbf{1}_G\right) = \mathbf{E}(Y \mathbf{1}_G) \tag{13.2.2}$$

for any $G \in \mathcal{G}$. If $\mathcal{G} = \sigma(X)$, then this definition reduces precisely to the previous one, with $\mathbf{P}(A \mid X)$ being shorthand for $\mathbf{P}(A \mid \sigma(X))$, and $\mathbf{E}(Y \mid X)$ being shorthand for $\mathbf{E}(Y \mid \sigma(X))$. Similarly, we shall write e.g. $\mathbf{E}(Y \mid X_1, X_2)$ as shorthand for $\mathbf{E}(Y \mid \sigma(X_1, X_2))$, where $\sigma(X_1, X_2) = \sigma(\{\{X_1 \leq a\} \cap \{X_2 \leq b\} : a, b \in \mathbf{R}\})$ is the σ-algebra generated by X_1 and X_2. As before, these conditional random variables exist (and are unique up to a set of probability 0) by the Radon-Nikodym Theorem.

Two further examples may help to clarify matters. If $\mathcal{G} = \{\emptyset, \Omega\}$ (i.e., $\mathcal{G}$ is the trivial σ-algebra), then $\mathbf{P}(A \mid \mathcal{G}) = \mathbf{P}(A)$ and $\mathbf{E}(Y \mid \mathcal{G}) = \mathbf{E}(Y)$ are *constants*. On the other hand, if $\mathcal{G} = \mathcal{F}$ (i.e., $\mathcal{G}$ is the full σ-algebra), or equivalently if $A \in \mathcal{G}$ and X is $\mathcal{G}$-measurable, then $\mathbf{P}(A \mid \mathcal{G}) = \mathbf{1}_A$ and $\mathbf{E}(X \mid \mathcal{G}) = X$. Intuitively, if $\mathcal{G}$ is small then the conditional values cannot depend too much on the sample point ω, but rather they must represent average values. On the other hand, if $\mathcal{G}$ is large then the conditional values can (and must) be very close approximations to the unconditional values. In brief, the *larger* $\mathcal{G}$ is, the *more* random are conditionals with respect to $\mathcal{G}$. See also Exercise 13.4.1.

Exercise 13.2.3. Let $\mathcal{G}_1$ and $\mathcal{G}_2$ be two sub-σ-algebras.
(a) Prove that if Z is $\mathcal{G}_1$-measurable, and $\mathcal{G}_1 \subseteq \mathcal{G}_2$, then Z is also $\mathcal{G}_2$-measurable.
(b) Prove that if Z is $\mathcal{G}$-measurable, and also Z is $\mathcal{G}'$-measurable, then Z is $(\mathcal{G} \cap \mathcal{G}')$-measurable.

Exercise 13.2.4.   Let S_1 and S_2 be two disjoint events, and let $\mathcal{G}$ be a sub-σ-algebra. Show that the following hold with probability 1. [Hint: Use proof by contradiction.]
(a) $0 \le \mathbf{P}(S_1 \,|\, \mathcal{G}) \le 1$.
(b) $\mathbf{P}(S_1 \overset{\bullet}{\cup} S_2 \,|\, \mathcal{G}) = \mathbf{P}(S_1 \,|\, \mathcal{G}) + \mathbf{P}(S_2 \,|\, \mathcal{G})$.

Remark 13.2.5.   Exercise 13.2.4 shows that conditional probabilities behave "essentially" like ordinary probabilities. Now, the additivity property (b) is only guaranteed to hold with probability 1, and the exceptional set of probability 0 could perhaps be different for each S_1 and S_2. This suggests that perhaps $P(S \,|\, \mathcal{G})$ cannot be defined in a consistent, countably additive way for all $S \in \mathcal{F}$. However, it is a fact that if a random variable Y is real-valued (or, more generally, takes values in a *Polish* space, i.e. a complete separable metric space like $\mathbf{R}$), then there always exist *regular conditional distributions*, which are versions of $\mathbf{P}(Y \in B \,|\, \mathcal{G})$ defined precisely (not just w.p. 1) for all Borel B in a countably additive way; see e.g. Theorem 10.2.2 of Dudley (1989), or Theorem 33.3 of Billingsley (1995).

We close with two final results about conditional expectations.

Proposition 13.2.6.   *Let X and Y be random variables, and let $\mathcal{G}$ be a sub-σ-algebra. Suppose that $\mathbf{E}(Y)$ and $\mathbf{E}(XY)$ are finite, and furthermore that X is $\mathcal{G}$-measurable. Then with probability 1,*

$$\mathbf{E}(XY \,|\, \mathcal{G}) \;=\; X\,\mathbf{E}(Y \,|\, \mathcal{G}).$$

That is, we can "factor" X out of the conditional expectation.

Proof.   Clearly $X\,\mathbf{E}(Y \,|\, \mathcal{G})$ is $\mathcal{G}$-measurable. Furthermore, if $X = 1_{G_0}$, with $G_0, G \in \mathcal{G}$, then using the definition of $\mathbf{E}(Y \,|\, \mathcal{G})$, we have

$$\mathbf{E}\left(X\,\mathbf{E}(Y \,|\, \mathcal{G})\,1_G\right) = \mathbf{E}\left(\mathbf{E}(Y \,|\, \mathcal{G})\,1_{G \cap G_0}\right) = \mathbf{E}\left(Y 1_{G \cap G_0}\right) = \mathbf{E}\left(XY 1_G\right),$$

so that (13.2.2) holds in this case. But then by the usual linearity and monotone convergence arguments, (13.2.2) holds for general X. It then follows from the definition that $X\,\mathbf{E}(Y \,|\, \mathcal{G})$ is indeed a version of $\mathbf{E}(XY \,|\, \mathcal{G})$. ∎

For our final result, suppose that $\mathcal{G}_1 \subseteq \mathcal{G}_2$. Note that since $\mathbf{E}(Y \,|\, \mathcal{G}_1)$ is $\mathcal{G}_1$-measurable, it is also $\mathcal{G}_2$-measurable, so from the above discussion we have $\mathbf{E}\left(\mathbf{E}(Y \,|\, \mathcal{G}_1) \,|\, \mathcal{G}_2\right) = \mathbf{E}(Y \,|\, \mathcal{G}_1)$. That is, conditioning first on $\mathcal{G}_1$ and then on $\mathcal{G}_2$ is equivalent to conditioning just on the smaller sub-σ-algebra, $\mathcal{G}_1$. What is perhaps surprising is that we obtain the same result if we condition in the opposite order:

Proposition 13.2.7. *Let Y be a random variable with finite mean, and let $\mathcal{G}_1 \subseteq \mathcal{G}_2$ be two sub-σ-algebras. Then with probability 1,*

$$\mathbf{E}\left(\mathbf{E}(Y \mid \mathcal{G}_2) \mid \mathcal{G}_1\right) = \mathbf{E}(Y \mid \mathcal{G}_1).$$

That is, conditioning first on $\mathcal{G}_2$ and then on $\mathcal{G}_1$ is equivalent to conditioning just on the smaller sub-σ-algebra $\mathcal{G}_1$.

Proof. By definition, $\mathbf{E}\left(\mathbf{E}(Y \mid \mathcal{G}_2) \mid \mathcal{G}_1\right)$ is $\mathcal{G}_1$-measurable. Hence, to show that it is a version of $\mathbf{E}(Y \mid \mathcal{G}_1)$, it suffices to check that, for any $G \in \mathcal{G}_1 \subseteq \mathcal{G}_2$, $\mathbf{E}\left[\mathbf{E}\left(\mathbf{E}(Y \mid \mathcal{G}_2) \mid \mathcal{G}_1\right) \mathbf{1}_G\right] = \mathbf{E}(Y \mathbf{1}_G)$. But using the definitions of $\mathbf{E}(\cdots \mid \mathcal{G}_1)$ and $\mathbf{E}(\cdots \mid \mathcal{G}_2)$, respectively, and recalling that $G \in \mathcal{G}_1$ and $G \in \mathcal{G}_2$, we have that

$$\mathbf{E}\left[\mathbf{E}\left(\mathbf{E}(Y \mid \mathcal{G}_2) \mid \mathcal{G}_1\right) \mathbf{1}_G\right] = \mathbf{E}\left(\mathbf{E}(Y \mid \mathcal{G}_2) \mathbf{1}_G\right) = \mathbf{E}(Y \mathbf{1}_G). \qquad \blacksquare$$

In the special case $\mathcal{G}_1 = \mathcal{G}_2 = \mathcal{G}$, we obtain that $\mathbf{E}[\mathbf{E}(Y \mid \mathcal{G}) \mid \mathcal{G}] = \mathbf{E}[Y \mid \mathcal{G}]$. In words, repeating the operation "take conditional expectation with respect to $\mathcal{G}$" multiple times is equivalent to doing it just once. That is, conditional expectation is a *projection operator*, and can be thought of as "projecting" the random variable Y onto the σ-algebra $\mathcal{G}$.

13.3. Conditional variance.

Given jointly defined random variables X and Y, we can define the *conditional variance* of Y given X by

$$\mathbf{Var}(Y \mid X) = \mathbf{E}[(Y - \mathbf{E}(Y \mid X))^2 \mid X].$$

Intuitively, $\mathbf{Var}(Y \mid X)$ is a measure of how much uncertainty there is in Y even after we know X. Since X may well provide some information about Y, we might expect that on average $\mathbf{Var}(Y \mid X) < \mathbf{Var}(Y)$. That is indeed the case. More precisely:

Theorem 13.3.1. *Let Y be a random variable, and $\mathcal{G}$ a sub-σ-algebra. If $\mathbf{Var}(Y) < \infty$, then*

$$\mathbf{Var}(Y) = \mathbf{E}[\mathbf{Var}(Y \mid \mathcal{G})] + \mathbf{Var}[\mathbf{E}(Y \mid \mathcal{G})].$$

Proof. We compute (writing $m = \mathbf{E}(Y) = \mathbf{E}[\mathbf{E}(Y \mid \mathcal{G})]$, and using (13.1.3) and Exercise 13.2.4) that

$$
\begin{aligned}
\mathbf{Var}(Y) &= \mathbf{E}[(Y - m)^2] \\
&= \mathbf{E}\left[\mathbf{E}[(Y - m)^2 \mid \mathcal{G}]\right] \\
&= \mathbf{E}\left[\mathbf{E}[(Y - \mathbf{E}(Y \mid \mathcal{G}) + \mathbf{E}(Y \mid \mathcal{G}) - m)^2 \mid \mathcal{G}]\right] \\
&= \mathbf{E}\left[\mathbf{E}[(Y - \mathbf{E}(Y \mid \mathcal{G}))^2 \mid \mathcal{G}] + \mathbf{E}[(\mathbf{E}(Y \mid \mathcal{G}) - m)^2 \mid \mathcal{G}]\right. \\
&\qquad \left. + 2\,\mathbf{E}[\mathbf{E}[(Y - \mathbf{E}(Y \mid \mathcal{G}))(\mathbf{E}(Y \mid \mathcal{G}) - m) \mid \mathcal{G}]\right] \\
&= \mathbf{E}\left[\mathbf{Var}(Y \mid \mathcal{G})\right] + \mathbf{Var}\left[\mathbf{E}(Y \mid \mathcal{G})\right] + 0,
\end{aligned}
$$

since using Proposition 13.2.6, we have

$$\mathbf{E}[(Y - \mathbf{E}(Y \mid \mathcal{G}))\,(\mathbf{E}(Y \mid \mathcal{G}) - m) \mid \mathcal{G}] \;=\; (\mathbf{E}(Y \mid \mathcal{G}) - m)\,\mathbf{E}[(Y - \mathbf{E}(Y \mid \mathcal{G})) \mid \mathcal{G}]$$

$$=\; (\mathbf{E}(Y \mid \mathcal{G}) - m)\,\big[\mathbf{E}(Y \mid \mathcal{G}) - \mathbf{E}(Y \mid \mathcal{G})\big] \;=\; 0. \qquad\blacksquare$$

For example, if $\mathcal{G} = \sigma(X)$, then $\mathbf{E}[\mathbf{Var}(Y \mid X)]$ represents the average uncertainty in Y once X is known, while $\mathbf{Var}[\mathbf{E}(Y \mid X)]$ represents the uncertainty in Y caused by uncertainty in X. Theorem 13.3.1 asserts that the total variance of Y is given by the sum of these two contributions.

13.4. Exercises.

Exercise 13.4.1. Let A and B be events, with $0 < \mathbf{P}(B) < 1$. Let $\mathcal{G} = \sigma(B)$ be the σ-algebra generated by B.
(a) Describe $\mathcal{G}$ explicitly.
(b) Compute $\mathbf{P}(A \mid \mathcal{G})$ explicitly.
(c) Relate $\mathbf{P}(A \mid \mathcal{G})$ to the earlier notion of $\mathbf{P}(A \mid B) = \mathbf{P}(A \cap B) / \mathbf{P}(B)$.
(d) Similarly, for a random variable Y with finite mean, compute $\mathbf{E}(Y \mid \mathcal{G})$, and relate it to the earlier notion of $\mathbf{E}(Y \mid B) = \mathbf{E}(Y\,\mathbf{1}_B) / \mathbf{P}(B)$.

Exercise 13.4.2. Let $\mathcal{G}$ be a sub-σ-algebra, and let A be any event. Define the random variable X to be the indicator function $\mathbf{1}_A$. Prove that $\mathbf{E}(X \mid \mathcal{G}) = \mathbf{P}(A \mid \mathcal{G})$ with probability 1.

Exercise 13.4.3. Suppose X and Y are discrete random variables. Let $q(x, y) = \mathbf{P}(X = x,\, Y = y)$.
(a) Show that with probability 1,

$$\mathbf{E}(Y \mid X) \;=\; \frac{\sum_y y\, q(X, y)}{\sum_z q(X, z)}\,.$$

[Hint: One approach is to first argue that it suffices in (13.1.6) to consider the case $S = \{x_0\}$.]
(b) Compute $\mathbf{P}(Y = y \mid X)$. [Hint: Use Exercise 13.4.2 and part (a).]
(c) Show that $\mathbf{E}(Y \mid X) = \sum_y y\, \mathbf{P}(Y = y \mid X)$.

Exercise 13.4.4. Let X and Y be random variables with joint distribution given by $\mathcal{L}(X, Y) = d\mathbf{P} = f(x, y)\,\lambda_2(dx, dy)$, where λ_2 is two-dimensional Lebesgue measure, and $f : \mathbf{R}^2 \to \mathbf{R}$ is a non-negative Borel-measurable function with $\int_{\mathbf{R}^2} f\, d\lambda_2 = 1$. (Example 13.1.1 corresponds to the case $f(x, y) = \frac{1}{2}\mathbf{1}_T(x, y)$.) Show that we can take $\mathbf{P}(Y \in B \mid X) = \int_B g_X(y)\lambda(dy)$ and $\mathbf{E}(Y \mid X) = \int_{\mathbf{R}} y g_X(y)\lambda(dy)$, where the function g_x :

$\mathbf{R} \to \mathbf{R}$ is defined by $g_x(y) = \frac{f(x,y)}{\int_{\mathbf{R}} f(x,t)\lambda(dt)}$ whenever $\int_{\mathbf{R}} f(x,t)\lambda(dt)$ is positive and finite, otherwise (say) $g_x(y) = 0$.

Exercise 13.4.5. Let $\Omega = \{1, 2, 3\}$, and define random variables X and Y by $Y(\omega) = \omega$, and $X(1) = X(2) = 5$ and $X(3) = 6$. Let $Z = \mathbf{E}(Y \mid X)$.
(a) Describe $\sigma(X)$ precisely.
(b) Describe (with proof) $Z(\omega)$ for each $\omega \in \Omega$.

Exercise 13.4.6. Let $\mathcal{G}$ be a sub-σ-algebra, and let X and Y be two independent random variables. Prove by example that $\mathbf{E}(X \mid \mathcal{G})$ and $\mathbf{E}(Y \mid \mathcal{G})$ need not be independent. [Hint: Don't forget Exercise 3.6.3(a).]

Exercise 13.4.7. Suppose Y is $\sigma(X)$-measurable, and also X and Y are independent. Prove that there is $C \in \mathbf{R}$ with $\mathbf{P}(Y = C) = 1$. [Hint: First prove that $\mathbf{P}(Y \le y) = 0$ or 1 for each $y \in \mathbf{R}$.]

Exercise 13.4.8. Suppose Y is $\mathcal{G}$-measurable. Prove that $\mathbf{Var}(Y \mid \mathcal{G}) = 0$.

Exercise 13.4.9. Suppose X and Y are independent.
(a) Prove that $\mathbf{E}(Y \mid X) = \mathbf{E}(Y)$ w.p. 1.
(b) Prove that $\mathbf{Var}(Y \mid X) = \mathbf{Var}(Y)$ w.p. 1.
(c) Explicitly verify Theorem 13.3.1 (with $\mathcal{G} = \sigma(X)$) in this case.

Exercise 13.4.10. Give an example of jointly defined random variables which are *not* independent, but such that $\mathbf{E}(Y \mid X) = \mathbf{E}(Y)$ w.p. 1.

Exercise 13.4.11. Let X and Y be jointly defined random variables.
(a) Suppose $\mathbf{E}(Y \mid X) = \mathbf{E}(Y)$ w.p. 1. Prove that $\mathbf{E}(XY) = \mathbf{E}(X)\mathbf{E}(Y)$.
(b) Give an example where $\mathbf{E}(XY) = \mathbf{E}(X)\mathbf{E}(Y)$, but it is *not* the case that $\mathbf{E}(Y \mid X) = \mathbf{E}(Y)$ w.p. 1.

Exercise 13.4.12. Let $\{Z_n\}$ be independent, each with finite mean. Let $X_0 = a$, and $X_n = a + Z_1 + \ldots + Z_n$ for $n \ge 1$. Prove that

$$\mathbf{E}(X_{n+1} \mid X_0, X_1, \ldots, X_n) = X_n + \mathbf{E}(Z_{n+1}).$$

Exercise 13.4.13. Let $X_0, X_1, \ldots$ be a Markov chain on a countable state space S, with transition probabilities $\{p_{ij}\}$, and with $\mathbf{E}|X_n| < \infty$ for all n. Prove that with probability 1:
(a) $\mathbf{E}(X_{n+1} \mid X_0, X_1, \ldots, X_n) = \sum_{j \in \mathcal{X}} j \, p_{X_n j}$. [Hint: Don't forget Exercise 13.4.3.]
(b) $\mathbf{E}(X_{n+1} \mid X_n) = \sum_{j \in \mathcal{X}} j \, p_{X_n j}$.

13.5. Section summary.

This section discussed conditioning, with an emphasis on the problems that arise (and their partial resolution) when conditioning on events of probability 0, and more generally on random variables and on sub-σ-algebras. It presented definitions, examples, and various properties of conditional probability and conditional expectation in these cases.

14. Martingales.

In this section we study a special kind of stochastic process called a martingale. A stochastic process $X_0, X_1, \ldots$ is a *martingale* if $\mathbf{E}|X_n| < \infty$ for all n, and with probability 1,

$$\mathbf{E}\left(X_{n+1} \mid X_0, X_1, \ldots, X_n\right) = X_n.$$

Of course, this really means that $\mathbf{E}\left(X_{n+1} \mid \mathcal{F}_n\right) = X_n$, where $\mathcal{F}_n$ is the σ-algebra $\sigma(X_0, X_1, \ldots, X_n)$. Intuitively, it says that *on average*, the value of X_{n+1} is the same as that of X_n.

Remark 14.0.1. More generally, we can define a martingale by specifying that $\mathbf{E}\left(X_{n+1} \mid \mathcal{F}_n\right) = X_n$ for *some* choice of increasing sub-σ-fields $\mathcal{F}_n$ such that X_n is measurable with respect to $\mathcal{F}_n$. But then $\sigma(X_0, \ldots, X_n) \subseteq \mathcal{F}_n$, so by Proposition 13.2.7,

$$\mathbf{E}(X_{n+1} \mid X_0, \ldots, X_n) = \mathbf{E}\big[\mathbf{E}(X_{n+1} \mid \mathcal{F}_n) \mid X_0, \ldots, X_n\big]$$

$$= \mathbf{E}[X_n \mid X_0, \ldots, X_n] = X_n,$$

so the above definition is also satisfied, i.e. the two definitions are equivalent.

Markov chains often provide good examples of martingales (though there are non-Markovian martingales too, cf. Exercise 14.4.1). Indeed, by Exercise 13.4.13, a Markov chain on a countable state space S, with transition probabilities p_{ij}, and with $\mathbf{E}|X_n| < \infty$, will be a martingale provided that

$$\sum_{j \in S} j\, p_{ij} = i, \qquad i \in S. \tag{14.0.2}$$

(Intuitively, given that the chain is at state i at time n, on average it will still equal i at time $n+1$.) An important specific case is simple symmetric random walk, where $S = \mathbf{Z}$ and $X_0 = 0$ and $p_{i,i-1} = p_{i,i+1} = \frac{1}{2}$ for all $i \in \mathbf{Z}$.

We shall also have occasion to consider submartingales and supermartingales. The sequence $X_0, X_1, \ldots$ is a *submartingale* if $\mathbf{E}|X_n| < \infty$ for all n, and also

$$\mathbf{E}\left(X_{n+1} \mid X_0, X_1, \ldots, X_n\right) \geq X_n. \tag{14.0.3}$$

It is a *supermartingale* if $\mathbf{E}|X_n| < \infty$ for all n, and also

$$\mathbf{E}\left(X_{n+1} \mid X_0, X_1, \ldots, X_n\right) \leq X_n.$$

(These names are very standard, even though they are arguably the reverse of what they should be.) Thus, a process is a martingale if and only if it is both a submartingale and a supermartingale. And, $\{X_n\}$ is a submartingale if and only if $\{-X_n\}$ is a supermartingale. Furthermore, again by Exercise 13.4.13, a Markov chain $\{X_n\}$ with $\mathbf{E}|X_n| < \infty$ is a submartingale if $\sum_{j \in S} j\, p_{ij} \geq i$ for all $i \in S$, or is a supermartingale if $\sum_{j \in S} j\, p_{ij} \leq i$ for all $i \in S$. Similarly, if $X_0 = 0$ and $X_n = Z_1 + \ldots + Z_n$, where $\{Z_i\}$ are i.i.d. with finite mean, then by Exercise 13.4.12, $\{X_n\}$ is a martingale if $\mathbf{E}(Z_i) = 0$, is a supermartingale if $\mathbf{E}(Z_i) \leq 0$; or is a submartingale if $\mathbf{E}(Z_i) \geq 0$.

If $\{X_n\}$ is a submartingale, then taking expectations of both sides of (14.0.3) gives that $\mathbf{E}(X_{n+1}) \geq \mathbf{E}(X_n)$, so by induction

$$\mathbf{E}(X_n) \geq \mathbf{E}(X_{n-1}) \geq \ldots \geq \mathbf{E}(X_1) \geq \mathbf{E}(X_0). \qquad (14.0.4)$$

Similarly, using Proposition 13.2.7,

$$\mathbf{E}\left[X_{n+2} \mid X_0, \ldots, X_n\right] = \mathbf{E}\left[\mathbf{E}(X_{n+2} \mid X_0, \ldots, X_{n+1}) \mid X_0, \ldots, X_n\right]$$

$$\geq \mathbf{E}\left[X_{n+1} \mid X_0, \ldots, X_n\right] \geq X_n,$$

so by induction

$$\mathbf{E}\left(X_m \mid X_0, X_1, \ldots, X_n\right) \geq X_n, \qquad m > n. \qquad (14.0.5)$$

For supermartingales, analogous statements to (14.0.4) and (14.0.5) follow with each $\geq$ replaced by $\leq$, and for martingales they can be replaced by $=$.

14.1. Stopping times.

If $X_0, X_1, \ldots$ is a martingale, then as in (14.0.4), $\mathbf{E}(X_n) = \mathbf{E}(X_0)$ for all $n \in \mathbf{N}$. But what about $\mathbf{E}(X_\tau)$, where τ is a *random* time? That is, if we define a new random variable Y by $Y(\omega) = X_{\tau(\omega)}(\omega)$, then must we also have $\mathbf{E}(Y) = \mathbf{E}(X_0)$? If τ is independent of $\{X_n\}$, then $\mathbf{E}(X_\tau)$ is simply a weighted average of different $\mathbf{E}(X_n)$, and therefore still equals $\mathbf{E}(X_0)$. But what if τ and $\{X_n\}$ are not independent?

We shall assume that τ is a *stopping time*, i.e. a non-negative-integer-valued random variable with the property that $\{\tau = n\} \in \sigma(X_0, \ldots, X_n)$. (Intuitively, this means that one can determine if $\tau = n$ just by knowing the values $X_0, \ldots, X_n$; τ does not "look into the future" to decide whether or not to stop at time n.) Under these conditions, must it be true that $\mathbf{E}(X_\tau) = \mathbf{E}(X_0)$?

The answer to this question is no in general. Indeed, consider simple symmetric random walk, with $X_0 = 0$, and let $\tau = \inf\{n \geq 0; X_n = -5\}$. Then τ is a stopping time, since $\{\tau = n\} = \{X_0 \neq -5, \ldots, X_{n-1} \neq -5, X_n = -5\} \in \sigma(X_0, \ldots, X_n)$. Furthermore, since $\{X_n\}$ is recurrent, we have $\mathbf{P}(\tau < \infty) = 1$. On the other hand, clearly $X_\tau = -5$ with probability 1, so that $\mathbf{E}(X_\tau) = -5$, not 0.

However, for *bounded* stopping times the situation is quite different, as the following result (one version of the *Optional Sampling Theorem*) shows.

Theorem 14.1.1. Let $\{X_n\}$ be a submartingale, let $M \in \mathbf{N}$, and let τ_1, τ_2 be stopping times such that $0 \leq \tau_1 \leq \tau_2 \leq M$ with probability 1. Then $\mathbf{E}(X_{\tau_2}) \geq \mathbf{E}(X_{\tau_1})$.

Proof. Note first that $\{\tau_1 < k \leq \tau_2\} = \{\tau_1 \leq k - 1\} \cap \{\tau_2 \leq k - 1\}^C$, so that (since τ_1 and τ_2 are stopping times) the event $\{\tau_1 < k \leq \tau_2\}$ is in $\sigma(X_0, \ldots, X_{k-1})$. We therefore have, using a telescoping series, linearity of expectation, and the definition of $\mathbf{E}(\cdots \mid X_0, \ldots, X_{k-1})$, that

$$
\begin{aligned}
&\mathbf{E}(X_{\tau_2}) - \mathbf{E}(X_{\tau_1}) \\
&= \mathbf{E}(X_{\tau_2} - X_{\tau_1}) \\
&= \mathbf{E}\left(\sum_{k=\tau_1+1}^{\tau_2}(X_k - X_{k-1})\right) \\
&= \mathbf{E}\left(\sum_{k=1}^{M}(X_k - X_{k-1})\mathbf{1}_{\tau_1 < k \leq \tau_2}\right) \\
&= \sum_{k=1}^{M}\mathbf{E}\left((X_k - X_{k-1})\mathbf{1}_{\tau_1 < k \leq \tau_2}\right) \\
&= \sum_{k=1}^{M}\mathbf{E}\left((\mathbf{E}(X_k \mid X_0, \ldots, X_{k-1}) - X_{k-1})\mathbf{1}_{\tau_1 < k \leq \tau_2}\right).
\end{aligned}
$$

But since $\{X_n\}$ is a submartingale, $(\mathbf{E}(X_k \mid X_0, \ldots, X_{k-1}) - X_{k-1}) \geq 0$. Therefore, $\mathbf{E}(X_{\tau_2}) - \mathbf{E}(X_{\tau_1}) \geq 0$, as claimed. ∎

Corollary 14.1.2. Let $\{X_n\}$ be a martingale, let $M \in \mathbf{N}$, and let τ_1, τ_2 be stopping times such that $0 \leq \tau_1 \leq \tau_2 \leq M$ with probability 1. Then $\mathbf{E}(X_{\tau_2}) = \mathbf{E}(X_{\tau_1})$.

Proof. Since both $\{X_n\}$ and $\{-X_n\}$ are submartingales, the proposition gives that $\mathbf{E}(X_{\tau_2}) \geq \mathbf{E}(X_{\tau_1})$ and also $\mathbf{E}(-X_{\tau_2}) \geq \mathbf{E}(-X_{\tau_1})$. The result follows. ∎

Setting $\tau_1 = 0$, we obtain:

Corollary 14.1.3. If $\{X_n\}$ be a martingale, and τ is a bounded stopping time, then $\mathbf{E}(X_\tau) = \mathbf{E}(X_0)$.

Example 14.1.4. Let $\{X_n\}$ be simple symmetric random walk with $X_0 = 0$, and let $\tau = \min(10^{12}, \inf\{n \geq 0; X_n = -5\})$. Then τ is indeed a bounded stopping time (with $\tau \leq 10^{12}$), so we must have $\mathbf{E}(X_\tau) = 0$. This is surprising since the probability is extremely high that $\tau < 10^{12}$, and in this case $X_\tau = -5$. However, in the rare case that $\tau = 10^{12}$, X_τ will (on average) be so large as to cancel this -5 out. More precisely, using the observation (13.0.1) that $\mathbf{E}(Z \,|\, B) = \mathbf{E}(Z\,\mathbf{1}_B)\,/\,\mathbf{P}(B)$ when $\mathbf{P}(B) > 0$, we have that

$$0 \;=\; \mathbf{E}(X_\tau) \;=\; \mathbf{E}\big(X_\tau \mathbf{1}_{\tau < 10^{12}}\big) + \mathbf{E}\big(X_\tau \mathbf{1}_{\tau = 10^{12}}\big)$$

$$= \; \mathbf{E}\big(X_\tau \,|\, \tau < 10^{12}\big)\,\mathbf{P}(\tau < 10^{12}) + \mathbf{E}\big(X_\tau \,|\, \tau = 10^{12}\big)\,\mathbf{P}(\tau = 10^{12})$$

$$= \; (-5)\,\mathbf{P}(\tau < 10^{12}) + \mathbf{E}\big(X_{10^{12}} \,|\, \tau = 10^{12}\big)\,\mathbf{P}(\tau = 10^{12})\,.$$

Here $\mathbf{P}(\tau < 10^{12}) \approx 1$ and $\mathbf{P}(\tau = 10^{12}) \approx 0$, but $\mathbf{E}\big(X_{10^{12}} \,|\, \tau = 10^{12}\big)$ is so huge that the equation still holds.

A generalisation of Theorem 14.1.1, allowing for unbounded stopping times, is as follows.

Theorem 14.1.5.   Let $\{X_n\}$ be a martingale with stopping time τ. Suppose $\mathbf{P}(\tau < \infty) = 1$, and $\mathbf{E}|X_\tau| < \infty$, and $\lim_{n\to\infty} \mathbf{E}[X_n \mathbf{1}_{\tau > n}] = 0$. Then $\mathbf{E}(X_\tau) = \mathbf{E}(X_0)$.

Proof. Let $Z_n = X_{\min(\tau,n)}$ for $n = 0, 1, 2, \ldots$. Then $Z_n = X_\tau \mathbf{1}_{\tau \le n} + X_n \mathbf{1}_{\tau > n} = X_\tau - X_\tau \mathbf{1}_{\tau > n} + X_n \mathbf{1}_{\tau > n}$, so $X_\tau = Z_n - X_n \mathbf{1}_{\tau > n} + X_\tau \mathbf{1}_{\tau > n}$. Hence,

$$\mathbf{E}(X_\tau) = \mathbf{E}(Z_n) - \mathbf{E}[X_n \mathbf{1}_{\tau > n}] + \mathbf{E}[X_\tau \mathbf{1}_{\tau > n}]\,.$$

Since $\min(\tau, n)$ is a bounded stopping time (cf. Exercise 14.4.6(b)), it follows from Corollary 14.1.3 that $\mathbf{E}(Z_n) = \mathbf{E}(X_0)$ for all n. As $n \to \infty$, the second term goes to 0 by assumption. Also, the third term goes to 0 by the Dominated Convergence Theorem, since $E|X_\tau| < \infty$, and $\mathbf{1}_{\tau > n} \to 0$ w.p. 1 since $\mathbf{P}[\tau < \infty] = 1$. Hence, letting $n \to \infty$, we obtain that $\mathbf{E}(X_\tau) = \mathbf{E}(X_0)$. ∎

Remark 14.1.6.   Theorem 14.1.5 in turn implies Corollary 14.1.3, since if $\tau \le M$, then $X_n \mathbf{1}_{\tau > n} = 0$ for all $n > M$, and also

$$\mathbf{E}|X_\tau| \;=\; \mathbf{E}|X_0 \mathbf{1}_{\tau=0} + \ldots + X_M \mathbf{1}_{\tau=M}| \;\le\; \mathbf{E}|X_0| + \ldots + \mathbf{E}|X_M| \;<\; \infty\,.$$

However, this reasoning is circular, since we used Corollary 14.1.3 in the proof of Theorem 14.1.5.

Corollary 14.1.7.   Let $\{X_n\}$ be a martingale with stopping time τ, such that $\mathbf{P}[\tau < \infty] = 1$. Assume that $\{X_n\}$ is bounded up to time τ, i.e. that there is $M < \infty$ such that $|X_n|\,\mathbf{1}_{n \le \tau} \le M\,\mathbf{1}_{n \le \tau}$ for all n. Then $\mathbf{E}(X_\tau) = \mathbf{E}(X_0)$.

Proof. We have $|X_\tau| \leq M$, so that $\mathbf{E}|X_\tau| \leq M < \infty$. Also $|\mathbf{E}(X_n \mathbf{1}_{\tau>n})| \leq \mathbf{E}(|X_n| \mathbf{1}_{\tau>n}) \leq \mathbf{E}(M \mathbf{1}_{\tau>n}) = M \mathbf{P}(\tau > n)$, which converges to 0 as $n \to \infty$ since $\mathbf{P}[\tau < \infty] = 1$. Hence, the result follows from Theorem 14.1.3. ∎

Remark 14.1.8. For submartingales $\{X_n\}$ we may replace $=$ by $\geq$ in Corollary 14.1.3 and Theorem 14.1.5 and Corollary 14.1.7, while for super-martingales we may replace $=$ by $\leq$.

Another approach is given by:

Theorem 14.1.9. Let τ be a stopping time for a martingale $\{X_n\}$, with $\mathbf{P}(\tau < \infty) = 1$. Then $\mathbf{E}(X_\tau) = \mathbf{E}(X_0)$ if and only if $\lim_{n\to\infty} \mathbf{E}[X_{\min(\tau,n)}] = \mathbf{E}[\lim_{n\to\infty} X_{\min(\tau,n)}]$.

Proof. By Corollary 14.1.3, $\lim_{n\to\infty} \mathbf{E}[X_{\min(\tau,n)}] = \lim_{n\to\infty} \mathbf{E}(X_0) = \mathbf{E}(X_0)$. Also, since $\mathbf{P}(\tau < \infty) = 1$, we must have $\mathbf{P}[\lim_{n\to\infty} X_{\min(\tau,n)} = X_\tau] = 1$, so $\mathbf{E}[\lim_{n\to\infty} X_{\min(\tau,n)}] = \mathbf{E}(X_\tau)$. The result follows. ∎

Combining Theorem 14.1.9 with the Dominated Convergence Theorem yields:

Corollary 14.1.10. Let τ be a stopping time for a martingale $\{X_n\}$, with $\mathbf{P}(\tau < \infty) = 1$. Suppose there is a random variable Y with $\mathbf{E}(Y) < \infty$ and $|X_{\min(\tau,n)}| \leq Y$ for all n. Then $\mathbf{E}(X_\tau) = \mathbf{E}(X_0)$.

As a particular case, we have:

Corollary 14.1.11. Let τ be a stopping time for a martingale $\{X_n\}$. Suppose $\mathbf{E}(\tau) < \infty$, and $|X_n - X_{n-1}| \leq M < \infty$ for all n. Then $\mathbf{E}(X_\tau) = \mathbf{E}(X_0)$.

Proof. Let $Y = |X_0| + M\tau$. Then $\mathbf{E}(Y) \leq \mathbf{E}|X_0| + M\mathbf{E}(\tau) < \infty$. Also,

$$|X_{\min(\tau,n)}| \leq |X_0| + M \min(\tau,n) \leq Y.$$

Hence, the result follows from Corollary 14.1.10. ∎

Similarly, combining Theorem 14.1.9 with the Uniform Integrability Convergence Theorem yields:

Corollary 14.1.12. Let τ be a stopping time for a martingale $\{X_n\}$, with $\mathbf{P}(\tau < \infty) = 1$. Suppose

$$\lim_{\alpha\to\infty} \sup_n \mathbf{E}\left(|X_{\min(\tau,n)}| \mathbf{1}_{|X_{\min(\tau,n)}| \geq \alpha}\right) = 0.$$

Then $\mathbf{E}(X_\tau) = \mathbf{E}(X_0)$.

Example 14.1.13. *(Sequence waiting time.)* Let $\{r_n\}_{n\geq 1}$ be infinite fair coin tossing (cf. Subsection 2.6), and let $\tau = \inf\{n \geq 3 : r_{n-2} = 1, r_{n-1} = 0, r_n = 1\}$ be the first time the sequence heads-tails-heads is completed. Surprisingly, $\mathbf{E}(\tau)$ can be computed using martingales. Indeed, suppose that at each time n, a new player appears and bets \$1 on heads, then if they win they bet \$2 on tails, then if they win again they bet \$4 on heads. (They stop betting as soon as they either lose once or win three bets in a row.) Let S_n be the total amount won by all the betters by time n. Then by construction, $\{S_n\}$ is a martingale with stopping time τ, and furthermore $|S_n - S_{n-1}| \leq 7 < \infty$. Also, $\mathbf{E}(\tau) < \infty$, and $S_\tau = -\tau + 10$ by Exercise 14.4.12. Hence, by Corollary 14.1.11, $0 = \mathbf{E}(S_\tau) = -\mathbf{E}(\tau) + 10$, whence $\mathbf{E}(\tau) = 10$. (See also Exercise 14.4.13.)

Finally, we observe the following fact about $\mathbf{E}(X_\tau)$ when $\{X_n\}$ is a general random walk, i.e. is given by sums of i.i.d. random variables. (If the $\{Z_i\}$ are bounded, then part (a) below also follows from applying Corollary 14.1.11 to $\{X_n - nm\}$.)

Theorem 14.1.14. *(Wald's theorem)* Let $\{Z_i\}$ be i.i.d. with finite mean m. Let $X_0 = a$, and $X_n = a + Z_1 + \ldots + Z_n$ for $n \geq 1$. Let τ be a stopping time for $\{X_n\}$, with $\mathbf{E}(\tau) < \infty$. Then
(a) $\mathbf{E}(X_\tau) = a + m\,\mathbf{E}(\tau)$; and furthermore
(b) if $m = 0$ and $\mathbf{E}(Z_1^2) < \infty$, then $\mathbf{Var}(X_\tau) = \mathbf{E}(Z_1^2)\,\mathbf{E}(\tau)$.

Proof. Since $\{i \leq \tau\} = \{\tau > i-1\} \in \sigma(X_0, \ldots, X_{i-1}) = \sigma(Z_0, \ldots, Z_{i-1})$, it follows from Lemma 3.5.2 that $\{i \leq \tau\}$ and $\{Z_i \in B\}$ are independent for any Borel $B \subseteq \mathbf{R}$, so that $\mathbf{1}_{i \leq \tau}$ and Z_i are independent for each i.

For part (a), assume first that the Z_i are non-negative. It follows from countable linearity and independence that

$$\mathbf{E}(X_\tau - a) = \mathbf{E}(Z_1 + \ldots + Z_\tau) = \mathbf{E}\left(\sum_{i=1}^{\infty} Z_i\,\mathbf{1}_{i\leq\tau}\right)$$

$$= \sum_{i=1}^{\infty} \mathbf{E}(Z_i\,\mathbf{1}_{i\leq\tau}) = \sum_{i=1}^{\infty} \mathbf{E}(Z_i)\,\mathbf{E}(\mathbf{1}_{i\leq\tau})$$

$$= \sum_{i=1}^{\infty} \mathbf{E}(Z_1)\,\mathbf{E}(\mathbf{1}_{i\leq\tau}) = \mathbf{E}(Z_1)\sum_{i=1}^{\infty} \mathbf{P}(\tau \geq i) = \mathbf{E}(Z_1)\,\mathbf{E}(\tau),$$

the last equality following from Proposition 4.2.9.

For general Z_i, the above shows that $\mathbf{E}\big(\sum_i |Z_i|\, \mathbf{1}_{i \leq \tau}\big) = \mathbf{E}|Z_1|\,\mathbf{E}(\tau) < \infty$. Hence, by Exercise 4.5.14(a) or Corollary 9.4.4, we can write

$$\mathbf{E}(X_\tau - a) = \mathbf{E}\left(\sum_i Z_i\, \mathbf{1}_{i \leq \tau}\right) = \sum_i \mathbf{E}(Z_i\, \mathbf{1}_{i \leq \tau}) = \mathbf{E}(Z_1)\,\mathbf{E}(\tau).$$

For part (b), if $m = 0$ then from the above $\mathbf{E}(X_\tau) = a$. Assume first that τ is bounded, i.e. $\tau \leq M < \infty$. Then using part (a),

$$\mathbf{Var}(X_\tau) = \mathbf{E}((X_\tau - a)^2) = \mathbf{E}\left(\left(\sum_{i=1}^{M} Z_i\, \mathbf{1}_{i \leq \tau}\right)^2\right)$$

$$= \mathbf{E}\left(\sum_{i=1}^{M} (Z_i)^2\, \mathbf{1}_{i \leq \tau}\right) + 0 = \mathbf{E}(Z_1^2)\,\mathbf{E}(\tau),$$

where the cross terms equal 0 since as before $\{j \leq \tau\} \in \sigma(Z_0, \ldots, Z_{j-1})$ so $Z_i\, \mathbf{1}_{j \leq \tau}$ is independent of Z_j, and hence

$$\mathbf{E}\left(\sum_{i < j} Z_i Z_j\, \mathbf{1}_{i \leq \tau}\, \mathbf{1}_{j \leq \tau}\right) = \mathbf{E}\left(\sum_{1 \leq i < j \leq M} Z_i Z_j\, \mathbf{1}_{j \leq \tau}\right)$$

$$= \sum_{1 \leq i < j \leq M} \mathbf{E}(Z_i Z_j\, \mathbf{1}_{j \leq \tau}) = \sum_{1 \leq i < j \leq M} \mathbf{E}(Z_i\, \mathbf{1}_{j \leq \tau})\,\mathbf{E}(Z_j) = 0.$$

If τ is not bounded, then let $\rho_k = \min(\tau, k)$ be corresponding bounded stopping times. The above gives that $\mathbf{Var}(X_{\rho_k}) = \mathbf{E}(Z_1^2)\,\mathbf{E}(\rho_k)$ for each k. As $k \to \infty$, $\mathbf{E}(\rho_k) \to \mathbf{E}(\tau)$ by the Monotone Convergence Theorem, so the result follows from Lemma 14.1.15 below. ∎

The final argument in the proof of Theorem 14.1.14 requires two technical lemmas involving L^2 theory:

Lemma 14.1.15. *In the proof of Theorem 14.1.14, $\lim_{k \to \infty} \mathbf{Var}(X_{\rho_k}) = \mathbf{Var}(X_\tau)$.*

Proof (optional). Clearly $X_{\rho_k} \to X_\tau$ a.s. Let $\|W\| = \sqrt{\mathbf{E}(W^2)}$ be the usual L^2 norm. Note first that for $m < n$, using Theorem 14.1.14(a) and since the cross terms again vanish, we have

$$\|X_{\rho_n} - X_{\rho_m}\|^2 = \mathbf{E}[(X_{\rho_n} - X_{\rho_m})^2] = \mathbf{E}[(Z_{\rho_m+1} + \ldots + Z_{\rho_n})^2]$$

$$= \mathbf{E}[Z_{\rho_m+1}^2 + \ldots + Z_{\rho_n}^2] + 0 = \mathbf{E}[Z_1^2 + \ldots + Z_{\rho_n}^2] - \mathbf{E}[Z_1^2 + \ldots + Z_{\rho_m}^2]$$

$$= \mathbf{E}(Z_1^2)\mathbf{E}(\rho_n) - \mathbf{E}(Z_1^2)\mathbf{E}(\rho_m) = \mathbf{E}(Z_1^2)\big[\mathbf{E}(\rho_n) - \mathbf{E}(\rho_m)\big]$$

which $\to \mathbf{E}(Z_1^2)\big[\mathbf{E}(\tau) - \mathbf{E}(\tau)\big] = 0$ as $n, m \to \infty$. This means that the sequence $\{X_{\rho_n}\}_{n=1}^\infty$ is a *Cauchy sequence in* L^2, and since L^2 is *complete* (see e.g. Dudley (1989), Theorem 5.2.1), we must have $\lim_{n\to\infty} \|X_{\rho_n} - Y\| = 0$ for some random variable Y. It then follows from Lemma 14.1.16 below (with $p = 2$) that $Y = X_\tau$ a.s., so that $\lim_{n\to\infty} \|X_{\rho_n} - X_\tau\| = 0$. The triangle inequality then gives that $\big|\,\|X_{\rho_n} - a\| - \|X_\tau - a\|\,\big| \le \|X_{\rho_n} - X_\tau\|$, so $\lim_{n\to\infty} \|X_{\rho_n} - a\| = \|X_\tau - a\|$. Hence, $\lim_{n\to\infty} \|X_{\rho_n} - a\|^2 = \|X_\tau - a\|^2$, i.e. $\lim_{n\to\infty} \mathbf{Var}(X_{\rho_n}^2) = \mathbf{Var}(X_\tau^2)$. ∎

Lemma 14.1.16.   If $\{Y_n\} \to Y$ in L^p, i.e. $\lim_{n\to\infty} \mathbf{E}(|Y_n - Y|^p) = 0$, where $1 \le p < \infty$, then there is a subsequence $\{Y_{n_k}\}$ with $Y_{n_k} \to Y$ a.s. In particular, if $\{Y_n\} \to Y$ in L^p, and also $\{Y_n\} \to Z$ a.s., then $Y = Z$ a.s.

Proof (optional).   Let $\{Y_{n_k}\}$ be any subsequence such that $\mathbf{E}(|Y_{n_k} - Y|^p) \le 4^{-k}$, and let $A_k = \{\omega \in \Omega : |Y_{n_k}(\omega) - Y(\omega)| \ge 2^{-k/p}\}$. Then

$$4^{-k} \ge \mathbf{E}\big(|Y_{n_k} - Y|^p\big) \ge \mathbf{E}\big(|Y_{n_k} - Y|^p \mathbf{1}_{A_k}\big)$$

$$\ge (2^{-k/p})^p \, \mathbf{P}(A_k) = 2^{-k}\,\mathbf{P}(A_k).$$

Hence, $\mathbf{P}(A_k) \le 2^{-k}$, so $\sum_k \mathbf{P}(A_k) = 1 < \infty$, so by the Borel-Cantelli Lemma, $\mathbf{P}(\limsup_k A_k) = 0$. For any $\epsilon > 0$, since $\{\omega \in \Omega : |Y_{n_k}(\omega) - Y(\omega)| \ge \epsilon\} \subseteq A_k$ for all $k \ge p\log_2(1/\epsilon)$, this shows that $\mathbf{P}(|Y_{n_k}(\omega) - Y(\omega)| \ge \epsilon \ i.o.) \le \mathbf{P}(\limsup_k A_k) = 0$. Lemma 5.2.1 then implies that $\{Y_{n_k}\} \to Y$ a.s. ∎

14.2. Martingale convergence.

If $\{X_n\}$ is a martingale (or a submartingale), will it converge almost surely to some (random) value?

Of course, the answer to this question in general is no. Indeed, simple symmetric random walk is a martingale, but it is recurrent, so it will forever oscillate between all the integers, without ever settling down anywhere.

On the other hand, consider the Markov chain on the non-negative integers with (say) $X_0 = 50$, and with transition probabilities given by $p_{ij} = \frac{1}{2i+1}$ for $0 \le j \le 2i$, with $p_{ij} = 0$ otherwise. (That is, if $X_n = i$, then X_{n+1} is uniformly distributed on the $2i + 1$ points $0, 1, \ldots, 2i$.) This Markov chain is a martingale. Clearly if it converges at all it must converge to 0; but does it?

This question is answered by:

Theorem 14.2.1.   *(Martingale Convergence Theorem) Let $\{X_n\}$ be a submartingale. Suppose that $\sup_n \mathbf{E}|X_n| < \infty$. Then there is a (finite) random variable X such that $X_n \to X$ almost surely (a.s.).*

We shall prove Theorem 14.2.1 below. We first note the following immediate corollary.

Corollary 14.2.2.   *Let $\{X_n\}$ be a martingale which is non-negative (or more generally satisfies that either $X_n \geq C$ for all n, or $X_n \leq C$ for all n, for some $C \in \mathbf{R}$). Then there is a (finite) random variable X such that $X_n \to X$ a.s.*

Proof.   If $\{X_n\}$ is a non-negative martingale, then $\mathbf{E}|X_n| = \mathbf{E}(X_n) = \mathbf{E}(X_0)$, so the result follows directly from Theorem 14.2.1.
    If $X_n \geq C$ [respectively $X_n \leq C$], then the result follows since $\{X_n - C\}$ [respectively $\{-X_n + C\}$] is a non-negative martingale.   ∎

It follows from this corollary that the above Markov chain example does indeed converge to 0 a.s.

For a second example, suppose $X_0 = 50$ and that $p_{ij} = \frac{1}{2\min(i,\,100-i)+1}$ for $|j - i| \leq \min(i, 100 - i)$. This Markov chain lives on $\{0, 1, 2, \ldots, 100\}$, at each stage jumping uniformly to one of the points within $\min(i, 100 - i)$ of its current position. By Corollary 14.2.2, $\{X_n\} \to X$ a.s. for some random variable X, and indeed it is easily seen that $\mathbf{P}(X = 0) = \mathbf{P}(X = 100) = \frac{1}{2}$.

For a third example, let $S = \{2^n; n \in \mathbf{Z}\}$, with a Markov chain on S having transition probabilities $p_{i,2i} = \frac{1}{3}$, $p_{i,\frac{i}{2}} = \frac{2}{3}$. This is again a martingale, and in fact it converges to 0 a.s. (even though it is unbounded).

For a fourth example, let $S = \mathbf{N}$ be the set of positive integers, with $X_0 = 1$. Let $p_{ii} = 1$ for i even, with $p_{i,i-1} = 2/3$ and $p_{i,i+2} = 1/3$ for i odd. Then it is easy to see that this Markov chain is a non-negative martingale, which converges a.s. to a random variable X having the property that $\mathbf{P}(X = i) > 0$ for every even non-negative integer i.

It remains to prove Theorem 14.2.1. We require the following lemma.

Lemma 14.2.3.   *(Upcrossing Lemma) Let $\{X_n\}$ be a submartingale. For $M \in \mathbf{N}$ and $\alpha < \beta$, let*

$$U_M^{\alpha,\beta} = \sup\{k;\ \exists\, a_1 < b_1 < a_2 < b_2 < a_3 < \ldots$$

$$< a_k < b_k \leq M;\ X_{a_i} \leq \alpha,\ X_{b_i} \geq \beta\}.$$

(Intuitively, $U_M^{\alpha,\beta}$ represents the number of times the process $\{X_n\}$ "up-crosses" the interval $[\alpha, \beta]$ by time M.) Then

$$\mathbf{E}(U_M^{\alpha,\beta}) \leq \mathbf{E}|X_M - X_0| / (\beta - \alpha).$$

Proof. By Exercise 14.4.2, the sequence $\{\max(X_n, \alpha)\}$ is also a sub-martingale, and it clearly has the same number of upcrossings of $[\alpha, \beta]$ as does $\{X_n\}$. Furthermore, $|\max(X_M, \alpha) - \max(X_0, \alpha)| \leq |X_M - X_0|$. Hence, for the remainder of the proof we can (and do) replace X_n by $\max(X_n, \alpha)$, i.e. we assume that $X_n \geq \alpha$ for all n.

Let $u_0 = v_0 = 0$, and iteratively define u_j and v_j for $j \geq 1$ by

$$u_j = \min\left(M, \inf\{k > v_{j-1}; X_k \leq \alpha\}\right);$$

$$v_j = \min\left(M, \inf\{k > u_j; X_k \geq \beta\}\right).$$

We necessarily have $v_M = M$, so that

$$\mathbf{E}(X_M) = \mathbf{E}(X_{v_M})$$

$$= \mathbf{E}(X_{v_M} - X_{u_M} + X_{u_M} - X_{v_{M-1}} + X_{v_{M-1}} - X_{u_{M-1}} + \ldots + X_{u_1} - X_0 + X_0)$$

$$= \mathbf{E}(X_0) + \mathbf{E}\left(\sum_{k=1}^{M}(X_{v_k} - X_{u_k})\right) + \sum_{k=1}^{M} \mathbf{E}\left(X_{u_k} - X_{v_{k-1}}\right).$$

Now, $\mathbf{E}(X_{u_k} - X_{v_{k-1}}) \geq 0$ by Theorem 14.1.1, since $\{X_n\}$ is a sub-martingale and $v_{k-1} \leq u_k \leq M$. Hence, $\sum_{k=1}^{M} \mathbf{E}(X_{u_k} - X_{v_{k-1}}) \geq 0$. (This is the subtlest part of the proof; since usually $X_{u_k} \leq \alpha$ and $X_{v_{k-1}} \geq \beta$, it is surprising that $\mathbf{E}(X_{u_k} - X_{v_{k-1}}) \geq 0$.)

Furthermore, we claim that

$$\mathbf{E}\left(\sum_{k=1}^{M}(X_{v_k} - X_{u_k})\right) \geq \mathbf{E}\left((\beta - \alpha)\, U_M^{\alpha,\beta}\right). \tag{14.2.4}$$

Indeed, each upcrossing contributes at least $\beta - \alpha$ to the sum. And, any "null cycle" where $u_k = v_k = M$ contributes nothing, since then $X_{v_k} - X_{u_k} = X_M - X_M = 0$. Finally, we may have one "incomplete cycle" where $u_k < M$ but $v_k = M$. However, since we are assuming that $X_n \geq \alpha$, we must have $X_{v_k} - X_{u_k} = X_M - X_{u_k} \geq \alpha - \alpha = 0$ in this case; that is, such an incomplete cycle could only *increase* the sum. This establishes (14.2.4).

We conclude that $\mathbf{E}(X_M) \geq \mathbf{E}(X_0) + (\beta - \alpha)\mathbf{E}\left(U_M^{\alpha,\beta}\right)$. Re-arranging,

$$(\beta - \alpha)\mathbf{E}\left(U_M^{\alpha,\beta}\right) \leq \mathbf{E}(X_M) - \mathbf{E}(X_0) \leq \mathbf{E}|X_M - X_0|,$$ which gives the result. ∎

Proof of Theorem 14.2.1.   Let $K = \sup_n \mathbf{E}\left(|X_n|\right) < \infty$. Note first that by Fatou's lemma,

$$\mathbf{E}\left(\liminf_n |X_n|\right) \leq \liminf_n \mathbf{E}|X_n| \leq K < \infty.$$

It follows that $\mathbf{P}\left(|X_n| \to \infty\right) = 0$, i.e. $\{X_n\}$ will *not* diverge to $\pm\infty$.
    Suppose now that $\mathbf{P}\left(\liminf X_n < \limsup X_n\right) > 0$. Then since

$$\left\{\liminf X_n < \limsup X_n\right\}$$

$$= \bigcup_{q \in \mathbf{Q}} \bigcup_{k \in \mathbf{N}} \left\{\liminf X_n < q < q + \frac{1}{k} < \limsup X_n\right\},$$

it follows from countable subadditivity that we can find $\alpha, \beta \in \mathbf{Q}$ with

$$\mathbf{P}\left(\liminf X_n < \alpha < \beta < \limsup X_n\right) > 0.$$

With $U_M^{\alpha,\beta}$ as in Lemma 14.2.3, this implies that $\mathbf{P}\left(\lim_{M\to\infty} U_M^{\alpha,\beta} = \infty\right) >$ 0, so $\mathbf{E}\left(\lim_{M\to\infty} U_M^{\alpha,\beta}\right) = \infty$. Then by the monotone convergence theorem, $\lim_{M\to\infty} \mathbf{E}(U_M^{\alpha,\beta}) = \mathbf{E}\left(\lim_{M\to\infty} U_M^{\alpha,\beta}\right) = \infty$. But Lemma 14.2.3 says that for all $M \in \mathbf{N}$, $\mathbf{E}(U_M^{\alpha,\beta}) \leq \frac{\mathbf{E}|X_M - X_0|}{\beta - \alpha} \leq \frac{2K}{\beta - \alpha}$, a contradiction.
    We conclude that $\mathbf{P}(\lim_{n\to\infty} |X_n| = \infty) = 0$ and also $\mathbf{P}(\liminf X_n < \limsup X_n) = 0$. Hence, we must have $\mathbf{P}(\lim_{n\to\infty} X_n$ exists and is finite$) = 1$, as claimed. ∎

14.3. Maximal inequality.

    Markov's inequality says that for $\alpha > 0$, $\mathbf{P}(X_0 \geq \alpha) \leq \mathbf{P}(|X_0| \geq \alpha) \leq \mathbf{E}|X_0| / \alpha$. Surprisingly, for a submartingale, the same inequality holds with X_0 replaced by $\max_{0 \leq i \leq n} X_i$, even though usually $(\max_{0 \leq i \leq n} X_i) > X_0$.

Theorem 14.3.1.   *(Martingale maximal inequality) If $\{X_n\}$ is a submartingale, then for all $\alpha > 0$,*

$$\mathbf{P}\left[\left(\max_{0 \leq i \leq n} X_i\right) \geq \alpha\right] \leq \frac{\mathbf{E}|X_n|}{\alpha}.$$

Proof.   Let A_k be the event $\{X_k \geq \alpha$, but $X_i < \alpha$ for $i < k\}$, i.e. the event that the process *first* reaches α at time k. And let $A = \overset{\bullet}{\bigcup}_{0 \leq k \leq n} A_k$ be the event that the process reaches α by time n.

We need to show that $\mathbf{P}(A) \leq \mathbf{E}|X_n| / \alpha$, i.e. that $\alpha \mathbf{P}(A) \leq \mathbf{E}|X_n|$. To that end, we compute (letting $\mathcal{F}_k = \sigma(X_0, X_1, \dots, X_k)$) that

$$
\begin{aligned}
\alpha \mathbf{P}(A) &= \sum_{k=0}^n \alpha \mathbf{P}(A_k) \qquad \text{since } \{A_k\} \text{ disjoint} \\
&= \sum_{k=0}^n \mathbf{E}\left(\alpha \mathbf{1}_{A_k}\right) \\
&\leq \sum_{k=0}^n \mathbf{E}\left(X_k \mathbf{1}_{A_k}\right) \qquad \text{since } X_k \geq \alpha \text{ on } A_k \\
&\leq \sum_{k=0}^n \mathbf{E}\left(\mathbf{E}(X_n \mid \mathcal{F}_k)\mathbf{1}_{A_k}\right) \qquad \text{by (14.0.5)} \\
&= \sum_{k=0}^n \mathbf{E}\left(X_n \mathbf{1}_{A_k}\right) \qquad \text{since } A_k \in \mathcal{F}_k \\
&= \mathbf{E}\left(X_n \mathbf{1}_A\right) \qquad \text{since } \{A_k\} \text{ disjoint} \\
&\leq \mathbf{E}\left(|X_n| \mathbf{1}_A\right) \\
&\leq \mathbf{E}|X_n|,
\end{aligned}
$$

as required.   ∎

Theorem 14.3.1 is clearly false if $\{X_n\}$ is not a submartingale. For example, if the X_i are i.i.d. equal to 0 or 2 each with probability $\frac{1}{2}$, and if $\alpha = 2$, then $\mathbf{P}\left[(\max_{0 \leq i \leq n} X_i) \geq \alpha\right] = 1 - (\frac{1}{2})^{n+1}$, which for $n \geq 1$ is not $\leq \mathbf{E}|X_n| / \alpha = \frac{1}{2}$.

If $\{X_n\}$ is in fact a non-negative martingale, then $\mathbf{E}|X_n| = \mathbf{E}(X_n) = \mathbf{E}(X_0)$, so the bound in Theorem 14.3.1 does not depend on n. Hence, letting $n \to \infty$ and using continuity of probabilities, we obtain:

Corollary 14.3.2. *If $\{X_n\}$ is a non-negative martingale, then for all $\alpha > 0$,*

$$
\mathbf{P}\left[\left(\sup_{0 \leq i < \infty} X_i\right) \geq \alpha\right] \leq \frac{\mathbf{E}(X_0)}{\alpha}.
$$

For example, consider the third Markov chain example above, where $S = \{2^n; n \in \mathbf{Z}\}$ and $p_{i,2i} = \frac{1}{3}$, $p_{i,\frac{i}{2}} = \frac{2}{3}$. If, say, $X_0 = 1$, then we obtain that $\mathbf{P}\left[(\sup_{0 \leq i < \infty} X_i) \geq 2\right] \leq \frac{1}{2}$, which is perhaps surprising. (This result also follows from applying (7.2.7) to the simple non-symmetric random walk $\{\log_2 X_n\}$.)

Remark 14.3.3. (Martingale Central Limit Theorem) If $X_n = Z_0 + \dots + Z_n$, where $\{Z_i\}$ are i.i.d. with mean 0 and variance 1, then we already know from the classical Central Limit Theorem that $\frac{X_n}{\sqrt{n}} \Rightarrow N(0, 1)$, i.e. that the properly normalised X_n converge weakly to a standard normal distribution. In fact, a similar result is true for more general martingales $\{X_n\}$. Let $\mathcal{F}_n = \sigma(X_0, X_1, \dots, X_n)$. Set $\sigma_0^2 = \mathbf{Var}(X_0)$, and for $n \geq 1$ set

$$
\sigma_n^2 = \mathbf{Var}(X_n \mid \mathcal{F}_{n-1}) = \mathbf{E}\left(X_n^2 - X_{n-1}^2 \mid \mathcal{F}_{n-1}\right).
$$

Then follows from induction that $\mathbf{Var}(X_n) = \sum_{k=0}^n \mathbf{E}\left(\sigma_k^2\right)$, and of course $\mathbf{E}(X_n) = \mathbf{E}(X_0)$ does not grow with n. Hence, if we set $v_t = \min\{n \geq$

0; $\sum_{k=0}^{n} \sigma_k^2 \geq t\}$, then for large t, the random variable $X_{v_t}/\sqrt{t}$ has mean close to 0 and variance close to 1. We then have $\frac{X_{v_t}}{\sqrt{t}} \Rightarrow N(0,1)$ as $t \to \infty$ under certain conditions, for example if $\sum_n \sigma_n^2 = \infty$ with probability one and the differences $X_n - X_{n-1}$ are uniformly bounded. For a proof see e.g. Billingley (1995, Theorem 35.11).

14.4. Exercises.

Exercise 14.4.1. Let $\{Z_i\}$ be i.i.d. with $\mathbf{P}(Z_i = 0) = \mathbf{P}(Z_i = 1) = 1/2$. Let $X_0 = 0$, $X_1 = 2Z_1 - 1$, and for $n \geq 2$, $X_n = X_{n-1} + (1 + Z_1 + ... + Z_{n-1})(2Z_n - 1)$. (Intuitively, this corresponds to wagering, at each time n, one dollar more than the number of previous victories, on a fair bet.)
(a) Prove that $\{X_n\}$ is a martingale.
(b) Prove that $\{X_n\}$ is *not* a Markov chain.

Exercise 14.4.2. Let $\{X_n\}$ be a submartingale, and let $a \in \mathbf{R}$. Let $Y_n = \max(X_n, a)$. Prove that $\{Y_n\}$ is also a submartingale. (Hint: Use Exercise 4.5.2.)

Exercise 14.4.3. Let $\{X_n\}$ be a non-negative submartingale. Let $Y_n = (X_n)^2$. Assuming $\mathbf{E}(Y_n) < \infty$ for all n, prove that $\{Y_n\}$ is also a submartingale.

Exercise 14.4.4. The *conditional Jensen's inequality* states that if ϕ is a convex function, then $\mathbf{E}\left(\phi(X) \mid \mathcal{G}\right) \geq \phi\left(\mathbf{E}(X \mid \mathcal{G})\right)$.
(a) Assuming this, prove that if $\{X_n\}$ is a submartingale, then so is $\{\phi(X_n)\}$ whenever ϕ is non-decreasing and convex with $\mathbf{E}|\phi(X_n)| < \infty$ for all n.
(b) Show that the conclusions of the two previous exercises follow from part (a).

Exercise 14.4.5. Let Z be a random variable on a probability triple $(\Omega, \mathcal{F}, \mathbf{P})$, and let $\mathcal{G}_0 \subseteq \mathcal{G}_1 \subseteq \ldots \subseteq \mathcal{F}$ be a nested sequence of sub-σ-algebras. Let $X_n = \mathbf{E}(Z \mid \mathcal{G}_n)$. (If we think of $\mathcal{G}_n$ as the amount of information we have available at time n, then X_n represents our best guess of the value Z at time n.) Prove that $X_0, X_1, \ldots$ is a martingale. [Hint: Use Proposition 13.2.7 to show that $\mathbf{E}(X_{n+1} \mid \mathcal{G}_n) = X_n$. Then use the fact that X_i is $\mathcal{G}_i$-measurable to prove that $\sigma(X_0, \ldots, X_n) \subseteq \mathcal{G}_n$.]

Exercise 14.4.6. Let $\{X_n\}$ be a stochastic process, let τ and ρ be two non-negative-integer-valued random variables, and let $m \in \mathbf{N}$.
(a) Prove that τ is a stopping time for $\{X_n\}$ if and only if $\{\tau \leq n\} \in \sigma(X_0, \ldots, X_n)$ for all $n \geq 0$.

(b) Prove that if τ is a stopping time, then so is $\min(\tau, m)$.
(c) Prove that if τ and ρ are stopping times for $\{X_n\}$, then so is $\min(\tau, \rho)$.

Exercise 14.4.7. Let $C \in \mathbf{R}$, and let $\{Z_i\}$ be an i.i.d. collection of random variables with $\mathbf{P}[Z_i = -1] = 3/4$ and $\mathbf{P}[Z_i = C] = 1/4$. Let $X_0 = 5$, and $X_n = 5 + Z_1 + Z_2 + \ldots + Z_n$ for $n \geq 1$.
(a) Find a value of C such that $\{X_n\}$ is a martingale.
(b) For this value of C, prove or disprove that there is a random variable X such that as $n \to \infty$, $X_n \to X$ with probability 1.
(c) For this value of C, prove or disprove that $\mathbf{P}[X_n = 0$ for some $n \in \mathbf{N}] = 1$.

Exercise 14.4.8. Let $\{X_n\}$ be simple symmetric random walk, with $X_0 = 0$. Let $\tau = \inf\{n \geq 5 : X_{n+1} = X_n + 1\}$ be the first time after 4 which is just before the chain increases. Let $\rho = \tau + 1$.
(a) Is τ a stopping time? Is ρ a stopping time?
(b) Use Theorem 14.1.5 to compute $\mathbf{E}(X_\rho)$.
(c) Use the result of part (b) to compute $\mathbf{E}(X_\tau)$. Why does this not contradict Theorem 14.1.5?

Exercise 14.4.9. Let $0 < a < c$ be integers, with $c \geq 3$. Let $\{X_n\}$ be simple symmetric random walk with $X_0 = a$, let $\sigma = \inf\{n \geq 1 : X_n = 0$ or $c\}$, and let $\rho = \sigma - 1$. Determine whether or not $\mathbf{E}(X_\sigma) = a$, and whether or not $\mathbf{E}(X_\rho) = a$. Relate the results to Corollary 14.1.7.

Exercise 14.4.10. Let $0 < a < c$ be integers. Let $\{X_n\}$ be simple symmetric random walk, started at $X_0 = a$. Let $\tau = \inf\{n \geq 1; X_n = 0$ or $c\}$.
(a) Prove that $\{X_n\}$ is a martingale.
(b) Prove that $\mathbf{E}(X_\tau) = a$. [Hint: Use Corollary 14.1.7.]
(c) Use this fact to derive an alternative proof of the gambler's ruin formula given in Section 7.2, for the case $p = 1/2$.

Exercise 14.4.11. Let $0 < p < 1$ with $p \neq 1/2$, and let $0 < a < c$ be integers. Let $\{X_n\}$ be simple random walk with parameter p, started at $X_0 = a$. Let $\tau = \inf\{n \geq 1; X_n = 0$ or $c\}$. Let $Z_n = ((1-p)/p)^{X_n}$ for $n = 0, 1, 2, \ldots$.
(a) Prove that $\{Z_n\}$ is a martingale.
(b) Prove that $\mathbf{E}(Z_\tau) = ((1-p)/p)^a$. [Hint: Use Corollary 14.1.7.]
(c) Use this fact to derive an alternative proof of the gambler's ruin formula given in Section 7.2, for the case $p \neq 1/2$.

Exercise 14.4.12. Let $\{S_n\}$ and τ be as in Example 14.1.13.
(a) Prove that $\mathbf{E}(\tau) < \infty$. [Hint: Show that $\mathbf{P}(\tau > 3m) \leq (7/8)^m$, and

use Proposition 4.2.9.]

(b) Prove that $S_\tau = -\tau + 10$. [Hint: By considering the τ different players one at a time, argue that $S_\tau = (\tau - 3)(-1) + 7 - 1 + 1$.]

Exercise 14.4.13.   Similar to Example 14.1.13, let $\{r_n\}_{n \geq 1}$ be infinite fair coin tossing, $\sigma = \inf\{n \geq 3 : r_{n-2} = 0, r_{n-1} = 1, r_n = 1\}$, and $\rho = \inf\{n \geq 4 : r_{n-3} = 0, r_{n-2} = 1, r_{n-1} = 0, r_n = 1\}$.
(a) Describe σ and ρ in plain English.
(b) Compute $\mathbf{E}(\sigma)$.
(c) Does $\mathbf{E}(\sigma) = \mathbf{E}(\tau)$, with τ as in Example 14.1.13? Why or why not?
(d) Compute $\mathbf{E}(\rho)$.

Exercise 14.4.14.   Why does the proof of Theorem 14.1.1 fail if $M = \infty$? [Hint: Exercise 4.5.14 may help.]

Exercise 14.4.15.   Modify the proof of Theorem 14.1.1 to show that if $\{X_n\}$ is a submartingale with $|X_{n+1} - X_n| \leq M < \infty$, and $\tau_1 \leq \tau_2$ are stopping times (not necessarily bounded) with $\mathbf{E}(\tau_2) < \infty$, then we still have $\mathbf{E}(X_{\tau_2}) \geq \mathbf{E}(X_{\tau_1})$. [Hint: Corollary 9.4.4 may help.] (Compare Corollary 14.1.11.)

Exercise 14.4.16.   Let $\{X_n\}$ be simple symmetric random walk, with $X_0 = 10$. Let $\tau = \min\{n \geq 1; \ X_n = 0\}$, and let $Y_n = X_{\min(n,\tau)}$. Determine (with explanation) whether each of the following statements is true or false.
(a) $\mathbf{E}(X_{200}) = 10$.
(b) $\mathbf{E}(Y_{200}) = 10$.
(c) $\mathbf{E}(X_\tau) = 10$.
(d) $\mathbf{E}(Y_\tau) = 10$.
(e) There is a random variable X such that $\{X_n\} \to X$ a.s.
(f) There is a random variable Y such that $\{Y_n\} \to Y$ a.s.

Exercise 14.4.17.   Let $\{X_n\}$ be simple symmetric random walk with $X_0 = 0$, and let $\tau = \inf\{n \geq 0; X_n = -5\}$.
(a) What is $\mathbf{E}(X_\tau)$ in this case?
(b) Why does this fact not contradict Wald's theorem part (a) (with $a = m = 0$)?

Exercise 14.4.18.   Let $0 < p < 1$ with $p \neq 1/2$, and let $0 < a < c$ be integers. Let $\{X_n\}$ be simple random walk with parameter p, started at $X_0 = a$. Let $\tau = \inf\{n \geq 1; \ X_n = 0 \text{ or } c\}$. (Thus, from the gambler's ruin solution of Subsection 7.2, $\mathbf{P}(X_\tau = c) = [((1-p)/p)^a - 1] / [((1-p)/p)^c - 1]$.)
(a) Compute $\mathbf{E}(X_\tau)$ by direct computation.
(b) Use Wald's theorem part (a) to compute $\mathbf{E}(\tau)$ in terms of $\mathbf{E}(X_\tau)$.

(c) Prove that the game's expected duration satisfies $\mathbf{E}(\tau) = (a - c[((1 - p)/p)^a - 1] / [((1 - p)/p)^c - 1]) / (1 - 2p)$.
(d) Show that the limit of $\mathbf{E}(\tau)$ as $p \to 1/2$ is equal to $a(c - a)$.

Exercise 14.4.19. Let $0 < a < c$ be integers, and let $\{X_n\}$ be simple *symmetric* random walk with $X_0 = a$. Let $\tau = \inf\{n \geq 1; X_n = 0 \text{ or } c\}$.
(a) Compute $\mathbf{Var}(X_\tau)$ by direct computation.
(b) Use Wald's theorem part (b) to compute $\mathbf{E}(\tau)$ in terms of $\mathbf{Var}(X_\tau)$.
(c) Prove that the game's expected duration satisfies $\mathbf{E}(\tau) = a(c - a)$.
(d) Relate this result to part (d) of the previous exercise.

Exercise 14.4.20. Let $\{X_n\}$ be a martingale with $|X_{n+1} - X_n| \leq 10$ for all n. Let $\tau = \inf\{n \geq 1 : |X_n| \geq 100\}$.
(a) Prove or disprove that this implies that $\mathbf{P}(\tau < \infty) = 1$.
(b) Prove or disprove that this implies there is a random variable X with $\{X_n\} \to X$ a.s.
(c) Prove or disprove that this implies that $\mathbf{P}[\tau < \infty$, or there is a random variable X with $\{X_n\} \to X] = 1$. [Hint: Let $Y_n = X_{\min(\tau,n)}$.]

Exercise 14.4.21. Let $Z_1, Z_2, \ldots$ to be independent, with

$$P\left(Z_i = \frac{2^i}{2^i - 1}\right) = \frac{2^i - 1}{2^i} \quad \text{and} \quad P(Z_i = -2^i) = \frac{1}{2^i}.$$

Let $X_0 = 0$, and $X_n = Z_1 + \ldots + Z_n$ for $n \geq 1$.
(a) Prove that $\{X_n\}$ is a martingale. [Hint: Don't forget (14.0.2).]
(b) Prove that $\mathbf{P}[Z_i > 1 \ a.a.] = 1$, i.e. that with probability 1, $Z_i > 1$ for all but finitely many i. [Hint: Don't forget the Borel-Cantelli Lemma.]
(c) Prove that $\mathbf{P}[\lim_{n \to \infty} X_n = \infty] = 1$. (Hence, even though $\{X_n\}$ is a martingale and thus represents a player's fortune in a "fair" game, it is still certain that the player's fortune will converge to $+\infty$.)
(d) Why does this result not contradict Corollary 14.2.2?
(e) Let $\tau = \inf\{n \geq 1 : X_n \leq 0\}$, and let $Y_n = X_{\min(\tau,n)}$. Prove that $\{Y_n\}$ is also a martingale, and that $\mathbf{P}[\lim_{n \to \infty} Y_n = \infty] > 0$. Why does this result not contradict Corollary 14.2.2?

14.5. Section summary.

This section provided a brief introduction to martingales, including submartingales, supermartingales, and stopping times. Some examples were given, mostly arising from Markov chains. Various versions were of the Optional Sampling Theorem were proved, giving conditions such that $\mathbf{E}(X_\tau) = \mathbf{E}(X_0)$. This together with the Upcrossing Lemma was then used to prove the important Martingale Convergence Theorem. Finally, the Martingale maximal inequality was presented.

15. General stochastic processes.

We end this text with a brief look at more general stochastic processes. We attempt to give an intuitive discussion of this area, without being overly careful about mathematical precision. (A full treatment of these processes would require another course, perhaps following one of the books in Subsection B.5.) In particular, in this section a number of results are stated without being proved, and a number of equations are derived in an intuitive and non-rigorous manner. All exercises are grouped by subsection, rather than at the end of the entire section.

To begin, we define a (completely general) *stochastic process* to be any collection $\{X_t;\ t \in T\}$ of random variables defined jointly on some probability triple. Here T can be any non-empty index set. If $T = \{0, 1, 2, \ldots\}$ then this corresponds to our usual discrete-time stochastic processes; if $T = \mathbf{R}^{\geq 0}$ is the non-negative real numbers, then this corresponds to a continuous-time stochastic process as discussed below.

15.1. Kolmogorov Existence Theorem.

As with all mathematical objects, a proper analysis of stochastic processes should begin with a proof that they *exist*. In two places in this text (Theorems 7.1.1 and 8.1.1), we have proved the existence of certain random variables, defined on certain underlying probability triples, having certain specified properties. The Kolmogorov Existence Theorem is a huge generalisation of these results, which allows us to define stochastic processes quite generally, as we now discuss.

Given a stochastic process $\{X_t;\ t \in T\}$, and $k \in \mathbf{N}$, and a finite collection $t_1, \ldots, t_k \in T$ of distinct index values, we define the Borel probability measure $\mu_{t_1 \ldots t_k}$ on $\mathbf{R}^k$ by

$$\mu_{t_1 \ldots t_k}(H) = \mathbf{P}\big((X_{t_1}, \ldots, X_{t_k}) \in H\big), \qquad H \subseteq \mathbf{R}^k \text{ Borel}.$$

The distributions $\{\mu_{t_1 \ldots t_k};\ k \in \mathbf{N},\ t_1, \ldots, t_k \in T \text{ distinct}\}$ are called the *finite-dimensional distributions* for the stochastic process $\{X_t;\ t \in T\}$.

These finite-dimensional distributions clearly satisfy two sorts of *consistency conditions*:

(C1) If $(s(1), s(2), \ldots, s(k))$ is any *permutation* of $(1, 2, \ldots, k)$ (meaning that $s : \{1, 2, \ldots, k\} \to \{1, 2, \ldots, k\}$ is one-to-one), then for distinct $t_1, \ldots, t_k \in T$, and any Borel $H_1, \ldots, H_k \subseteq \mathbf{R}$, we have

$$\mu_{t_1 \ldots t_k}(H_1 \times \ldots \times H_k) = \mu_{t_{s(1)} \ldots t_{s(k)}}(H_{s(1)} \times \ldots \times H_{s(k)}). \quad (15.1.1)$$

That is, if we permute the indices t_i, and correspondingly modify the set $H = H_1 \times \ldots \times H_k$, then we do not change the probabilities. For example, we must have $\mathbf{P}(X \in A, \ Y \in B) = \mathbf{P}(Y \in B, \ X \in A)$, even though this will *not* usually equal $\mathbf{P}(Y \in A, \ X \in B)$.

(C2) For distinct $t_1, \ldots, t_k \in T$, and any Borel $H_1, \ldots, H_{k-1} \subseteq \mathbf{R}$, we have

$$\mu_{t_1 \ldots t_k}(H_1 \times \ldots \times H_{k-1} \times \mathbf{R}) = \mu_{t_1 \ldots t_{k-1}}(H_1 \times \ldots \times H_{k-1}). \quad (15.1.2)$$

That is, allowing X_{t_k} to be anywhere in $\mathbf{R}$ is equivalent to not mentioning X_{t_k} at all. For example, $\mathbf{P}(X \in A, \ Y \in \mathbf{R}) = \mathbf{P}(X \in A)$.

Conditions (C1) and (C2) are quite obvious and uninteresting. However, what is surprising is that they have an immediate *converse*; that is, for any collection of finite-dimensional distributions satisfying them, there exists a corresponding stochastic process. A formal statement is:

Theorem 15.1.3. *(Kolmogorov Existence Theorem) A family of Borel probability measures $\{\mu_{t_1 \ldots t_k}; \ k \in \mathbf{N}, \ t_i \in T \text{ distinct}\}$, with $\mu_{t_1 \ldots t_k}$ a measure on $\mathbf{R}^k$, satisfies the consistency conditions (C1) and (C2) above if and only if there exists a probability triple $(\mathbf{R}^T, \mathcal{F}^T, \mathbf{P})$, and random variables $\{X_t\}_{t \in T}$ defined on this triple, such that for all $k \in \mathbf{N}$, distinct $t_1, \ldots, t_k \in T$, and Borel $H \subseteq \mathbf{R}^k$, we have*

$$\mathbf{P}\left((X_{t_1}, \ldots, X_{t_k}) \in H\right) = \mu_{t_1 \ldots t_k}(H). \quad (15.1.4)$$

The theorem thus says that, under extremely general conditions, stochastic processes exist. Theorems 7.1.1 and 8.1.1 follow immediately as special cases (Exercises 15.1.6 and 15.1.7).

The "only if" direction is immediate, as discussed above. To prove the "if" direction, we can take

$$\mathbf{R}^T = \{\text{all functions } T \to \mathbf{R}\}$$

and

$$\mathcal{F}^T = \sigma\left\{\{X_t \in H\}; \ t \in T, \ H \subseteq \mathbf{R} \text{ Borel}\right\}.$$

The idea of the proof is to first use (15.1.4) to define $\mathbf{P}$ for subsets of the form $\{(X_{t_1}, \ldots, X_{t_k}) \in H\}$ (for distinct $t_1, \ldots, t_k \in T$, and Borel $H \subseteq \mathbf{R}^k$), and then extend the definition of $\mathbf{P}$ to the entire σ-algebra $\mathcal{F}^T$, analogously to the proof of the Extension Theorem (Theorem 2.3.1). The argument is somewhat involved; for details see e.g. Billingsley (1995, Theorem 36.1).

Exercise 15.1.5. Suppose that (15.1.1) holds whenever s is a *transposition*, i.e. a one-to-one function $s : \{1, 2, \ldots, k\} \to \{1, 2, \ldots, k\}$ such that

$s(i) = i$ for $k-2$ choices of $i \in \{1, 2, \ldots, k\}$. Prove that the first consistency condition is satisfied, i.e. that (15.1.1) holds for all permutations s.

Exercise 15.1.6. Let $X_1, X_2, \ldots$ be independent, with $X_n \sim \mu_n$.
(a) Specify the finite-dimensional distributions $\mu_{t_1, t_2, \ldots, t_k}$ for distinct non-negative integers $t_1, t_2, \ldots, t_k$.
(b) Prove that these $\mu_{t_1, t_2, \ldots, t_k}$ satisfy (15.1.1) and (15.1.2).
(c) Prove that Theorem 7.1.1 follows from Theorem 15.1.3.

Exercise 15.1.7. Consider the definition of a discrete Markov chain $\{X_n\}$, from Section 8.
(a) Specify the finite-dimensional distributions $\mu_{t_1, t_2, \ldots, t_k}$ for non-negative integers $t_1 < t_2 < \ldots < t_k$.
(b) Prove that they satisfy (15.1.1).
(c) Specify the finite-dimensional distributions $\mu_{t_1, t_2, \ldots, t_k}$ for general distinct non-negative integers $t_1, t_2, \ldots, t_k$, and explain why they satisfy (15.1.2). [Hint: It may be helpful to define the *order statistics*, whereby $t_{(r)}$ is the rth-largest element of $\{t_1, t_2, \ldots, t_k\}$ for $1 \leq r \leq k$.]
(d) Prove that Theorem 8.1.1 follows from Theorem 15.1.3.

15.2. Markov chains on general state spaces.

In Section 8 we considered Markov chains on countable state spaces S, in terms of an initial distribution $\{\nu_i\}_{i \in S}$ and transition probabilities $\{p_{ij}\}_{i,j \in S}$. We now generalise many of the notions there to general (perhaps uncountable) state spaces.

We require a general *state space* $\mathcal{X}$, which is any non-empty (perhaps uncountable) set, together with a σ-algebra $\mathcal{F}$ of measurable subsets. The *transition probabilities* are then given by $\{P(x, A)\}_{x \in \mathcal{X}, A \in \mathcal{F}}$. We make the following two assumptions:
(A1) For each fixed $x \in \mathcal{X}$, $P(x, \cdot)$ is a probability measure on $(\mathcal{X}, \mathcal{F})$.
(A2) For each fixed $A \in \mathcal{F}$, $P(x, A)$ is a measurable function of $x \in \mathcal{X}$.
Intuitively, $P(x, A)$ is the probability, if the chain is at a point x, that it will jump to the subset A at the next step. If $\mathcal{X}$ is countable, then $P(x, \{i\})$ corresponds to the transition probability p_{xi} of the discrete Markov chains of Section 8. But on a general state space, we may have $P(x, \{i\}) = 0$ for all $i \in \mathcal{X}$.

We also require an *initial distribution* ν, which is any probability distribution on $(\mathcal{X}, \mathcal{F})$. The transition probabilities and initial distribution then give rise to a (discrete-time, general state space, time-homogeneous) *Markov chain* $X_0, X_1, X_2, \ldots$, where

$$\mathbf{P}(X_0 \in A_0, \ X_1 \in A_1, \ldots, X_n \in A_n)$$

$$= \int_{x_0 \in A_0} \nu(dx_0) \int_{x_1 \in A_1} P(x_0, dx_1) \cdots$$

$$\cdots \int_{x_{n-1} \in A_{n-1}} P(x_{n-2}, dx_{n-1}) \, P(x_{n-1}, A_n). \qquad (15.2.1)$$

Note that these integrals (i.e., expected values) are well-defined because of condition (A2) above.

As before, we shall write $\mathbf{P}_x(\cdots)$ for the probability of an event conditional on $X_0 = x$, i.e. under the assumption that the initial distribution ν is a point-mass at the point x. And, we define higher-order transition probabilities inductively by $P^1(x, A) = P(x, A)$, and $P^{n+1}(x, A) = \int_{\mathcal{X}} P(x, dz) P^n(z, A)$ for $n \geq 1$.

Analogous to the countable state space case, we define a *stationary distribution* for a Markov chain to be a probability measure $\pi(\cdot)$ on $(\mathcal{X}, \mathcal{F})$, such that $\pi(A) = \int_{\mathcal{X}} \pi(dx) P(x, A)$ for all $A \in \mathcal{F}$. (This generalises our earlier definition $\pi_j = \sum_{i \in S} \pi_i \, p_{ij}$.) As in the countably infinite case, Markov chains on general state spaces may or may not have stationary distributions.

Example 15.2.2. Consider the Markov chain on the real line (i.e. with $\mathcal{X} = \mathbf{R}$), where $P(x, \cdot) = N(\frac{x}{2}, \frac{3}{4})$ for each $x \in \mathcal{X}$. In words, if X_n is equal to some real number x, then the conditional distribution of X_{n+1} will be normal, with mean $\frac{x}{2}$ and variance $\frac{3}{4}$. Equivalently, $X_{n+1} = \frac{1}{2}X_n + U_{n+1}$, where $\{U_n\}$ are i.i.d. with $U_n \sim N(0, \frac{3}{4})$. This example is analysed in Exercise 15.2.5 below; in particular, it is shown that $\pi(\cdot) = N(0, 1)$ is a stationary distribution for this chain.

For countable state spaces S, we defined irreducibility to mean that for all $i, j \in S$, there is $n \in \mathbf{N}$ with $p_{ij}^{(n)} > 0$. On uncountable state spaces this definition is of limited use, since we will often (e.g. in the above example) have $p_{ij}^{(n)} = 0$ for all $i, j \in \mathcal{X}$ and all $n \geq 1$. Instead, we say that a Markov chain on a general state space $\mathcal{X}$ is ϕ-*irreducible* if there is a non-zero, σ-finite measure ψ on $(\mathcal{X}, \mathcal{F})$ such that for any $A \in \mathcal{F}$ with $\psi(A) > 0$, we have $\mathbf{P}_x(\tau_A < \infty) > 0$ for all $x \in \mathcal{X}$. (Here $\tau_A = \inf\{n \geq 0; \ X_n \in A\}$ is the *first hitting time* of the subset A; thus, $\tau_A < \infty$ is the event that the chain eventually hits the subset A, and ϕ-irreducibility is the statement that the chain has positive probability of eventually hitting any subset A of positive ψ measure.)

Similarly, on countable state spaces S we defined aperiodicity to mean that for all $i \in S$, $\gcd\{n \geq 1 : p_{ii}^{(n)} > 0\} = 1$, but on uncountable state spaces we will usually have $p_{ii}^{(n)} = 0$ for all n. Instead, we define the *period* of a general-state-space Markov chain to be the largest (finite) positive integer d such that there are non-empty disjoint subsets $\mathcal{X}_1, \ldots, \mathcal{X}_d \subseteq \mathcal{X}$, with $P(x, \mathcal{X}_{i+1}) = 1$ for all $x \in \mathcal{X}_i$ $(1 \leq i \leq d - 1)$ and $P(x, \mathcal{X}_1) = 1$ for all $x \in \mathcal{X}_d$. The chain is *periodic* if its period is greater than 1, otherwise the chain is *aperiodic*.

In terms of these definitions, a fundamental theorem about general state space Markov chains is the following generalisation of Theorem 8.3.10. For a proof see e.g. Meyn and Tweedie (1993).

Theorem 15.2.3. *If a discrete-time Markov chain on a general state space is ϕ-irreducible and aperiodic, and furthermore has a stationary distribution $\pi(\cdot)$, then for π-almost every $x \in \mathcal{X}$, we have that*

$$\lim_{n \to \infty} \sup_{A \in \mathcal{F}} |P^n(x, A) - \pi(A)| \to 0.$$

In words, the Markov chain converges to its stationary distribution in the total variation distance metric.

Exercise 15.2.4. Consider a Markov chain which is ϕ-irreducible with respect to some non-zero σ-finite measure ψ, and which is periodic with corresponding disjoint subsets $\mathcal{X}_1, \ldots, \mathcal{X}_d$. Let $B = \bigcup_i \mathcal{X}_i$.
(a) Prove that $P^n(x, B^C) = 0$ for all $x \in B$.
(b) Prove that $\psi(B^C) = 0$.
(c) Prove that $\psi(\mathcal{X}_i) > 0$ for some i.

Exercise 15.2.5. Consider the Markov chain of Example 15.2.2.
(a) Let $\pi(\cdot) = N(0, 1)$ be the standard normal distribution. Prove that $\pi(\cdot)$ is a stationary distribution for this Markov chain.
(b) Prove that this Markov chain is ϕ-irreducible with respect to λ, where λ is Lebesgue measure on $\mathbf{R}$.
(c) Prove that this Markov chain is aperiodic. [Hint: Don't forget Exercise 15.2.4.]
(d) Apply Theorem 15.2.3 to this Markov chain, writing your conclusion as explicitly as possible.

Exercise 15.2.6. **(a)** Prove that a Markov chain on a countable state space $\mathcal{X}$ is ϕ-irreducible if and only if there is $j \in \mathcal{X}$ such that $\mathbf{P}_i(\tau_j < \infty) > 0$ for all $i \in \mathcal{X}$.
(b) Give an example of a Markov chain on a countable state space which is ϕ-irreducible, but which is *not* irreducible in the sense of Subsection 8.2.

Exercise 15.2.7. Show that, for a Markov chain on a countable state space S, the definition of aperiodicity from this subsection agrees with the previous definition from Definition 8.3.4.

Exercise 15.2.8. Consider a discrete-time Markov chain with state space $\mathcal{X} = \mathbf{R}$, and with transition probabilities such that $P(x, \cdot)$ is uniform on the interval $[x - 1, \, x + 1]$. Determine whether or not this chain is ϕ-irreducible.

Exercise 15.2.9. Recall that *counting measure* is the measure $\psi(\cdot)$ defined by $\psi(A) = |A|$, i.e. $\psi(A)$ is the number of elements in the set A, with $\psi(A) = \infty$ if the set A is infinite.
(a) For a Markov chain on a countable state space $\mathcal{X}$, prove that irreducibility in the sense of Subsection 8.2 is equivalent to ϕ-irreducibility with respect to counting measure on $\mathcal{X}$.
(b) Prove that the Markov chain of Exercise 15.2.5 is *not* ϕ-irreducible with respect to counting measure on $\mathbf{R}$. (That is, prove that there is a set A, and $x \in \mathcal{X}$, such that $\mathbf{P}_x(\tau_A < \infty) = 0$, even though $\psi(A) > 0$ where ψ is counting measure.)
(c) Prove that counting measure on $\mathbf{R}$ is *not* σ-finite (cf. Remark 4.4.3).

Exercise 15.2.10. Consider the Markov chain with $\mathcal{X} = \mathbf{R}$, and with $P(x, \cdot) = N(x, 1)$ for each $x \in \mathcal{X}$.
(a) Prove that this chain is ϕ-irreducible and aperiodic.
(b) Prove that this chain does *not* have a stationary distribution. Relate this to Theorem 15.2.3.

Exercise 15.2.11. Let $\mathcal{X} = \{1, 2, \ldots\}$. Let $P(1, \{1\}) = 1$, and for $x \geq 2$, $P(x, \{1\}) = 1/x^2$ and $P(x, \{x + 1\}) = 1 - (1/x^2)$.
(a) Prove that this chain has stationary distribution $\pi(\cdot) = \delta_1(\cdot)$.
(b) Prove that this chain is ϕ-irreducible and aperiodic.
(c) Prove that if $X_0 = x \geq 2$, then $\mathbf{P}[X_n = x + n \text{ for all } n] > 0$, and $\|P^n(x, \cdot) - \pi(\cdot)\| \not\to 0$. Relate this to Theorem 15.2.3.

Exercise 15.2.12. Show that the finite-dimensional distributions implied by (15.2.1) satisfy the two consistency conditions of the Kolmogorov Existence Theorem. What does this allow us to conclude?

15.3. Continuous-time Markov processes.

In general, a *continuous-time stochastic process* is a collection $\{X_t\}_{t \geq 0}$ of random variables, defined jointly on some probability triple, taking values in some state space $\mathcal{X}$ with σ-algebra $\mathcal{F}$, and indexed by the non-negative real numbers $T = \{t \geq 0\}$. Usually we think of the variable t as representing (continuous) time, so that X_t is the (random) state at some time $t \geq 0$.

Such a process is a (continuous-time, time-homogeneous) *Markov process* if there are transition probabilities $P^t(x, \cdot)$ for all $t \geq 0$ and all $x \in \mathcal{X}$, and an *initial distribution* ν, such that

$$\mathbf{P}(X_0 \in A_0, X_{t_1} \in A_1, \ldots, X_{t_n} \in A_n)$$

$$= \int_{x_0 \in A_0} \int_{x_{t_1} \in A_1} \cdots \int_{x_{t_n} \in A_n} \nu(dx_0)$$

$$P^{t_1}(x_0, dx_{t_1}) \, P^{t_2 - t_1}(x_{t_1}, dx_{t_2}) \ldots P^{t_n - t_{n-1}}(x_{t_{n-1}}, dx_{t_n}) \quad (15.3.1)$$

for all times $0 \leq t_1 < \cdots < t_n$ and all subsets $A_1, \ldots, A_n \in \mathcal{F}$. Letting $P^0(x, \cdot)$ be a point-mass at x, it then follows that

$$P^{s+t}(x, A) = \int P^s(x, dy) \, P^t(y, A), \quad s, t \geq 0, \ x \in \mathcal{X}, \ A \in \mathcal{F}. \quad (15.3.2)$$

(This is the *semigroup property* of the transition probabilities: $P^{s+t} = P^s \, P^t$.) On a countable state space $\mathcal{X}$, we sometimes write p_{ij}^t for $P^t(i, \{j\})$; (15.3.2) can then be written as $p_{ij}^{s+t} = \sum_{k \in \mathcal{X}} p_{ik}^t p_{kj}^s$. As in (8.0.5), $p_{ij}^0 = \delta_{ij}$, (i.e., at time $t = 0$, p_{ij}^0 equals 1 for $i = j$ and 0 otherwise).

Exercise 15.3.3. Let $\{P^t(x, \cdot)\}$ be a collection of Markov transition probabilities satisfying (15.3.2). Define finite-dimensional distributions $\mu_{t_1 \ldots t_k}$ for $t_1, \ldots, t_k \geq 0$ by $\mu_{t_1 \ldots t_k}(A_0 \times \ldots \times A_k) = \mathbf{P}(X_{t_1} \in A_1, \ldots, X_{t_k} \in A_k)$ as implied by (15.3.1).
(a) Prove that the Kolmogorov consistency conditions are satisfied by $\{\mu_{t_1 \ldots t_k}\}$. [Hint: You will need to use (15.3.2).]
(b) Apply the Kolmogorov Existence Theorem to $\{\mu_{t_1 \ldots t_k}\}$, and describe precisely the conclusion.

Another important concept for continuous-time Markov processes is the *generator*. If the state space $\mathcal{X}$ is countable (discrete), then the generator is a matrix $Q = (q_{ij})$, defined for $i, j \in \mathcal{X}$ by

$$q_{ij} = \lim_{t \searrow 0} \frac{p_{ij}^t - \delta_{ij}}{t}. \quad (15.3.4)$$

Since $0 \leq p_{ij}^t \leq 1$ for all $t \geq 0$ and all i and j, we see immediately that $q_{ii} \leq 0$, while $q_{ij} \geq 0$ if $i \neq j$. Also, since $\sum_j (p_{ij}^t - \delta_{ij}) = 1 - 1 = 0$, we have $\sum_j q_{ij} = 0$ for each i. The matrix Q represents a sort of "derivative" of P^t with respect to t. Hence, $p_{ij}^s \approx \delta_{ij} + s \, q_{ij}$ for small $s \geq 0$, i.e. $P^s = I + sQ + o(s)$ as $s \searrow 0$.

Exercise 15.3.5. Let $\{P^t\} = \{p_{ij}^t\}$ be the transition probabilities for a continuous-time Markov process on a finite state space $\mathcal{X}$, with finite generator $Q = \{q_{ij}\}$ given by (15.3.4). Show that $P^t = \exp(tQ)$. (Given a matrix M, we define $\exp(M)$ by $\exp(M) = I + M + \frac{1}{2}M^2 + \frac{1}{3!}M^3 + \dots$, where I is the identity matrix.) [Hint: $P^t = (P^{t/n})^n$, and as $n \to \infty$, we have $P^{t/n} = I + (t/n)Q + O(1/n^2)$.]

Exercise 15.3.6. Let $\mathcal{X} = \{1, 2\}$, and let $Q = (q_{ij})$ be the generator of a continuous-time Markov process on $\mathcal{X}$, with

$$Q = \begin{pmatrix} -3 & 3 \\ 6 & -6 \end{pmatrix}.$$

Compute the corresponding transition probabilities $P^t = (p_{ij}^t)$ of the process, for any $t > 0$. [Hint: Use the previous exercise. It may help to know that the eigenvalues of Q are 0 and -9, with corresponding left eigenvectors $(2, 1)$ and $(1, -1)$.]

The generator Q may be interpreted as giving "jump rates" for our Markov processes. If the chain starts at state i and next jumps to state j, then the time before it jumps is exponentially distributed with mean $1/q_{ij}$. We say that the process jumps from i to j at rate q_{ij}. Exercise 15.3.5 shows that the generator Q contains all the information necessary to completely reconstruct the transition probabilities P^t of the Markov process.

A probability distribution $\pi(\cdot)$ on $\mathcal{X}$ is a *stationary distribution* for the process $\{X_t\}_{t \geq 0}$ if $\int \pi(dx) P^t(x, \cdot) = \pi(\cdot)$ for all $t > 0$; if $\mathcal{X}$ is countable we can write this as $\pi_j = \sum_{i \in \mathcal{X}} \pi_i p_{ij}^t$ for all $j \in \mathcal{X}$ and all $t > 0$. Note that it is no longer sufficient to check this equation for just one particular value of t (e.g. $t = 1$). Similarly, we say that $\{X_t\}_{t \geq 0}$ is *reversible* with respect to $\pi(\cdot)$ if $\pi(dx) P^t(x, dy) = \pi(dy) P^t(y, dx)$ for all $x, y \in \mathcal{X}$ and all $t > 0$.

Exercise 15.3.7. Let $\{X_t\}_{t \geq 0}$ be a Markov process. Prove that
(a) if $\{X_t\}_{t \geq 0}$ is reversible with respect to $\pi(\cdot)$, then $\pi(\cdot)$ is a stationary distribution for $\{X_t\}_{t \geq 0}$.
(b) if for some $T > 0$ we have $\int \pi(dx) P^t(x, \cdot) = \pi(\cdot)$ for all $0 < t < T$, then $\pi(\cdot)$ is a stationary distribution for $\{X_t\}_{t \geq 0}$.
(c) if for some $T > 0$ we have $\pi(dx) P^t(x, dy) = \pi(dy) P^t(y, dx)$ for all $x, y \in \mathcal{X}$ and all $0 < t < T$, then $\{X_t\}_{t \geq 0}$ is reversible with respect to $\pi(\cdot)$.

Exercise 15.3.8. For a Markov chain on a finite state space $\mathcal{X}$ with generator Q, prove that $\{\pi_i\}_{i \in \mathcal{X}}$ is a stationary distribution if and only if $\pi Q = 0$, i.e. if and only if $\sum_{i \in \mathcal{X}} \pi_i q_{ij} = 0$ for all $j \in \mathcal{X}$.

Exercise 15.3.9. Let $\{Z_n\}$ be i.i.d. $\sim \mathbf{Exp}(5)$, so $\mathbf{P}[Z_n \geq z] = e^{-5z}$ for $z \geq 0$. Let $T_n = Z_1 + Z_2 + \dots + Z_n$ for $n \geq 1$. Let $\{X_t\}_{t \geq 0}$ be a continuous-time Markov process on the state space $\mathcal{X} = \{1, 2\}$, defined as follows. The process does not move except at the times T_n, and at each time T_n, the process jumps from its current state to the opposite state (i.e., from 1 to 2, or from 2 to 1). Compute the generator Q for this process. [Hint: You may use without proof the following facts, which are consequences of the "memoryless property" of the exponential distribution: for any $0 \le a \le b$, (i) $\mathbf{P}(\exists n : a < T_n \le b) = 1 - e^{-5(b-a)}$, which is $\approx 5(b-a)$ as $b \searrow a$; and also (ii) $\mathbf{P}(\exists n : a < T_n \le T_{n+1} \le b) = 1 - e^{-5(b-a)}(1 + 5(b-a))$, which is $\approx 25(b-a)^2$ as $b \searrow a$.]

Exercise 15.3.10. *(Poisson process.)* Let $\lambda > 0$, let $\{Z_n\}$ be i.i.d. $\sim \mathbf{Exp}(\lambda)$, and let $T_n = Z_1 + Z_2 + \ldots + Z_n$ for $n \ge 1$. Let $\{N_t\}_{t \ge 0}$ be a continuous-time Markov process on the state space $\mathcal{X} = \{0, 1, 2, \ldots\}$, with $N_0 = 0$, which does not move except at the times T_n, and increases by 1 at each time T_n. (Equivalently, $N_t = \#\{n \in \mathbf{N} : T_n \le t\}$; intuitively, N_t counts the number of events by time t.)
(a) Compute the generator Q for this process. [Don't forget the hint from the previous exercise.]
(b) Prove that $\mathbf{P}(N_t \le m) = e^{-\lambda t}(\lambda t)^m / m! + \mathbf{P}(N_t \le m - 1)$ for $m = 0, 1, 2, \ldots$. [Hint: First argue that $\mathbf{P}(N_t \le m) = \mathbf{P}(T_{m+1} > t)$ and that $T_m \sim \mathbf{Gamma}(m, \lambda)$, and then use integration by parts; you may assume the result and hints of Exercise 9.5.17.]
(c) Conclude that $\mathbf{P}(N_t = j) = e^{-\lambda t}(\lambda t)^j / j!$, i.e. that $N_t \sim \mathbf{Poisson}(\lambda t)$. [Remark: More generally, there are Poisson processes with variable rate functions $\lambda : [0, \infty) \to [0, \infty)$, for which $N_t \sim \mathbf{Poisson}\left(\int_0^t \lambda(s)\, ds\right)$.]

Exercise 15.3.11. Let $\{N_t\}_{t \ge 0}$ be a Poisson process with rate $\lambda > 0$.
(a) For $0 < s < t$, compute $\mathbf{P}(N_s = 1 \,|\, N_t = 1)$.
(b) Use this to specify precisely the conditional distribution of the first event time T_1, conditional on $N_t = 1$.
(c) More generally, for $r < t$ and $m \in \mathbf{N}$, specify the conditional distribution of T_m, conditional on $N_r = m - 1$ and $N_t = m$.

Exercise 15.3.12. (a) Let $\{N_t\}_{t \ge 0}$ be a Poisson process with rate $\lambda > 0$, let $0 < s < t$, and let U_1, U_2 be i.i.d. $\sim \mathbf{Uniform}[0, t]$.
(a) Compute $\mathbf{P}(N_s = 0 \,|\, N_t = 2)$.
(b) Compute $\mathbf{P}(U_1 > s \text{ and } U_2 > s)$.
(c) Compare the answers to parts (a) and (b). What does this comparison seem to imply?

15.4. Brownian motion as a limit.

Having discussed continuous-time processes on countable state spaces, we now turn to continuous-time processes on *continuous* state spaces. Such processes can move in continuous curves and therefore give rise to interesting *random functions*. The most important such process is *Brownian motion*, also called the *Wiener process*. Brownian motion is best understood as a limit of discrete time processes, as follows.

Let $Z_1, Z_2, \ldots$ be any sequence of i.i.d. bounded random variables having mean 0 and variance 1, e.g. $Z_i = \pm 1$ with probability $1/2$ each. For each $n \in \mathbf{N}$, define a discrete-time random process $\{Y_{\frac{i}{n}}^{(n)}\}_{i=0}^{\infty}$ iteratively by $Y_0^{(n)} = 0$, and

$$Y_{\frac{i+1}{n}}^{(n)} = Y_{\frac{i}{n}}^{(n)} + \frac{1}{\sqrt{n}}Z_{i+1}, \qquad i = 0, 1, 2, \ldots . \tag{15.4.1}$$

Thus, $Y_{\frac{i}{n}}^{(n)} = \frac{1}{\sqrt{n}}(Z_1 + Z_2 + \ldots + Z_i)$. In particular, $\{Y_{\frac{i}{n}}^{(n)}\}$ is like simple symmetric random walk, except with time sped up by a factor of n, and space shrunk by a factor of $\sqrt{n}$.

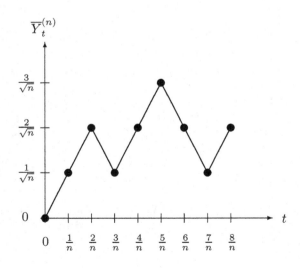

Figure 15.4.2. Constructing Brownian motion.

We can then "fill in" the missing values by linear interpolation on each interval $[\frac{i}{n}, \frac{i+1}{n}]$. In this way, we obtain a continuous-time process $\{\overline{Y}_t^{(n)}\}_{t\geq 0}$;

see Figure 15.4.2. Thus, $\{\overline{Y}_t^{(n)}\}_{t\geq 0}$ is a continuous-time process which agrees with $Y_t^{(n)}$ whenever $t = \frac{i}{n}$. Furthermore, $\overline{Y}_t^{(n)}$ is always within $O(1/n)$ of $Y_{\frac{\lfloor tn \rfloor}{n}}^{(n)}$, where $\lfloor r \rfloor$ is the greatest integer not exceeding r. Intuitively, then, $\{\overline{Y}_t^{(n)}\}_{t\geq 0}$ is like an ordinary discrete-time random walk, except that it has been interpolated to a continuous-time process, and furthermore time has been sped up by a factor of n and shrunk by a factor of $\sqrt{n}$. In summary, this process takes lots and lots of very small steps.

Remark. It is not essential that we choose the missing values so as to make the function *linear* on $[\frac{i}{n}, \frac{i+1}{n}]$. The same intuitive limit will be obtained no matter how the missing values are defined, provided only that they are close (as $n \to \infty$) to the corresponding $Y_{\frac{i}{n}}$ values. For example, another option is to use a *constant* interpolation of the missing values, by setting $\overline{Y}_t^{(n)} = Y_{\frac{\lfloor tn \rfloor}{n}}^{(n)}$. In fact, this constant interpolation would make the following calculations cleaner by eliminating the $O(1/n)$ errors. However, the linear interpolation has the advantage that each $\overline{Y}_t^{(n)}$ is then a *continuous* function of t, which makes it more intuitive why B_t is also continuous.

Now, the factors n and $\sqrt{n}$ have been chosen carefully. We see that $\overline{Y}_t^{(n)} \approx Y_{\frac{\lfloor tn \rfloor}{n}}^{(n)} = \frac{1}{\sqrt{n}}(Z_1 + Z_2 + \ldots + Z_{\lfloor tn \rfloor})$. Thus, $\overline{Y}_t^{(n)}$ is essentially $\frac{1}{\sqrt{n}}$ times the sum of $\lfloor tn \rfloor$ different i.i.d. random variables, each having mean 0 and variance 1. It follows from the ordinary Central Limit Theorem that, as $n \to \infty$ with t fixed, we have $\mathcal{L}\left(\overline{Y}_t^{(n)}\right) \Rightarrow N(0, t)$, i.e. for large n the random variable $\overline{Y}_t^{(n)}$ is approximately normal with mean 0 and variance t.

Let us note a couple of other facts. Firstly, for $s < t$, $\mathbf{E}\left(\overline{Y}_s^{(n)}\overline{Y}_t^{(n)}\right) \approx \frac{1}{n}\mathbf{E}\left(Y_{\frac{\lfloor sn \rfloor}{n}}^{(n)} Y_{\frac{\lfloor tn \rfloor}{n}}^{(n)}\right) = \mathbf{E}\left((Z_1 + \ldots + Z_{\lfloor sn \rfloor})(Z_1 + \ldots + Z_{\lfloor tn \rfloor})\right)$. Since the Z_i have mean 0 and variance 1, and are independent, this is equal to $\frac{\lfloor ns \rfloor}{n}$, which converges to s as $n \to \infty$.

Secondly, if $n \to \infty$ with $s < t$ fixed, the difference $\overline{Y}_t^{(n)} - \overline{Y}_s^{(n)}$ is within $O(1/n)$ of $Z_{\lfloor sn \rfloor + 1} + \ldots + Z_{\lfloor tn \rfloor}$. Hence, by the Central Limit Theorem, $\overline{Y}_t^{(n)} - \overline{Y}_s^{(n)}$ converges weakly to $N(0, t - s)$ as $n \to \infty$, and is in fact *independent* of $\overline{Y}_s^{(n)}$.

Brownian motion is, intuitively, the limit as $n \to \infty$ of the processes $\{\overline{Y}_t^{(n)}\}_{t\geq 0}$. That is, we can intuitively define Brownian motion $\{B_t\}_{t\geq 0}$ by saying that $B_t = \lim_{n\to\infty} \overline{Y}_t^{(n)}$ for each fixed $t \geq 0$. This analogy is very useful, in that all the $n \to \infty$ properties of $\overline{Y}_t^{(n)}$ discussed above will apply to Brownian motion. In particular:

- *normally distributed*: $\mathcal{L}(B_t) = N(0, t)$ for any $t \geq 0$.

- *covariance structure*: $\mathbf{E}(B_s B_t) = s$ for $0 \leq s \leq t$.

- *independent normal increments*: $\mathcal{L}(B_{t_2} - B_{t_1}) = N(0, t_2 - t_1)$, and $B_{t_4} - B_{t_3}$ is independent of $B_{t_2} - B_{t_1}$ whenever $0 \leq t_1 \leq t_2 \leq t_3 \leq t_4$.

In addition, Brownian motion has one other very important property. It has *continuous sample paths*, meaning that for each fixed $\omega \in \Omega$, the function $B_t(\omega)$ is a continuous function of t. (However, it turns out that, with probability one, Brownian motion is not *differentiable* anywhere at all!)

Exercise 15.4.3.   Let $\{B_t\}_{t \geq 0}$ be Brownian motion. Compute $\mathbf{E}[(B_2 + B_3 + 1)^2]$.

Exercise 15.4.4.   Let $\{B_t\}_{t \geq 0}$ be Brownian motion, and let $X_t = 2t + 3B_t$ for $t \geq 0$.
(a) Compute the distribution of X_t for $t \geq 0$.
(b) Compute $\mathbf{E}[(X_t)^2]$ for $t \geq 0$.
(c) Compute $\mathbf{E}(X_s X_t)$ for $0 < s < t$.

Exercise 15.4.5.   Let $\{B_t\}_{t \geq 0}$ be Brownian motion. Compute the distribution of Z_n for $n \in \mathbf{N}$, where $Z_n = \frac{1}{n}(B_1 + B_2 + \ldots + B_n)$.

Exercise 15.4.6.   **(a)**   Let $f : \mathbf{R} \to \mathbf{R}$ be a *Lipschitz function*, i.e. a function for which there exists $\alpha \in \mathbf{R}$ such that $|f(x) - f(y)| \leq \alpha |x - y|$ for all $x, y \in \mathbf{R}$. Compute $\lim_{h \searrow 0}(f(t+h) - f(t))^2/h$ for any $t \in \mathbf{R}$.
(b) Let $\{B_t\}$ be Brownian motion. Compute $\lim_{h \searrow 0} \mathbf{E}\left((B_{t+h} - B_t)^2/h\right)$ for any $t > 0$.
(c) What do parts (a) and (b) seem to imply about Brownian motion?

15.5. Existence of Brownian motion.

We thus see that Brownian motion has many useful properties. However, we cannot make mathematical use of these properties until we have established that Brownian motion *exists*, i.e. that there is a probability triple $(\Omega, \mathcal{F}, \mathbf{P})$, with random variables $\{B_t\}_{t \geq 0}$ defined on it, which satisfy the properties specified above.

Now, the Kolmogorov Existence Theorem (Theorem 15.1.3) is of some help here. It ensures the existence of a stochastic process having the same finite-dimensional distributions as our desired process $\{B_t\}$. At first glance this might appear to be all we need. Indeed, the properties of being normally distributed, of having the right covariance structure, and of having independent increments, all follow immediately from the finite-dimensional distributions.

The problem, however, is that finite-dimensional distributions cannot guarantee the property of *continuous sample paths*. To see this, let $\{B_t\}_{t\geq 0}$ be Brownian motion, and let U be a random variable which is distributed according to (say) Lebesgue measure on $[0,1]$. Define a new stochastic process $\{B'_t\}_{t\geq 0}$ by $B'_t = B_t + \mathbf{1}_{\{U=t\}}$. That is, B'_t is equal to B_t, except that if we happen to have $U = t$ then we add one to B_t. Now, since $\mathbf{P}(U = t) = 0$ for any fixed t, we see that $\{B'_t\}_{t\geq 0}$ has exactly the same finite-dimensional distributions as $\{B_t\}_{t\geq 0}$ does. On the other hand, obviously if $\{B_t\}_{t\geq 0}$ has continuous sample paths, then $\{B'_t\}_{t\geq 0}$ cannot. This shows that the Kolmogorov Existence Theorem is not sufficient to properly construct Brownian motion. Instead, one of several alternative approaches must be taken.

In the first approach, we begin by setting $B_0 = 0$ and choosing $B_1 \sim N(0,1)$. We then define $B_{\frac{1}{2}}$ to have its correct *conditional* distribution, conditional on the values of B_0 and B_1. Continuing, we then define $B_{\frac{1}{4}}$ to have its correct conditional distribution, conditional on the values of B_0 and $B_{\frac{1}{2}}$; and similarly define $B_{\frac{3}{4}}$ conditional on the values of $B_{\frac{1}{2}}$ and B_1. Iteratively, we see that we can define B_t for all values of t which are *dyadic rationals*, i.e. which are of the form $\frac{i}{2^n}$ for some integers i and n. We then argue that the function $\{B_t;\ t$ a non-negative dyadic rational$\}$ is *uniformly continuous*, and use this to argue that it can be "filled in" to a continuous function $\{B_t;\ t \geq 0\}$ defined for all non-negative real values t. For the details of this approach, see e.g. Billingsley (1995, Theorem 37.1).

In another approach, we define the processes $\{\overline{Y}_t^{(n)}\}$ as above, and observe that $\overline{Y}_t^{(n)}$ is a (random) continuous function of t. We then prove that as $n \to \infty$, the distributions of the functions $\{\overline{Y}_t^{(n)}\}_{t\geq 0}$ (regarded as probability distributions on the space of all continuous functions) are *tight*, and in fact converge to the distribution of $\{B_t\}_{t\geq 0}$ that we want. We then use this fact to conclude the existence of the random variables $\{B_t\}_{t\geq 0}$. For details of this approach, see e.g. Fristedt and Gray (1997, Section 19.2).

Exercise 15.5.1.   Construct a process $\{B''_t\}_{t\geq 0}$ which has the same finite-dimensional distributions as does $\{B_t\}_{t\geq 0}$ and $\{B'_t\}_{t\geq 0}$, but such that neither $\{B''_t\}_{t\geq 0}$ nor $\{B''_t - B'_t\}_{t\geq 0}$ has continuous sample paths.

Exercise 15.5.2.   Let $\{X_t\}_{t\in T}$ and $\{X'_t\}_{t\in T}$ be a stochastic processes with the *countable* time index T. Suppose $\{X_t\}_{t\in T}$ and $\{X'_t\}_{t\in T}$ have identical finite-dimensional distributions. Prove or disprove that $\{X_t\}_{t\in T}$ and $\{X'_t\}_{t\in T}$ must have the same full joint distribution.

15.6. Diffusions and stochastic integrals.

Suppose we are given continuous functions $\mu, \sigma : \mathbf{R} \to \mathbf{R}$, and we replace equation (15.4.1) by

$$Y_{\frac{i+1}{n}}^{(n)} = Y_{\frac{i}{n}}^{(n)} + \frac{1}{\sqrt{n}} Z_{i+1} \sigma \left(Y_{\frac{i}{n}}^{(n)} \right) + \frac{1}{n} \mu \left(Y_{\frac{i}{n}}^{(n)} \right) . \qquad (15.6.1)$$

That is, we multiply the effect of Z_{i+1} by $\sigma(Y_{\frac{i}{n}}^{(n)})$, and we add in an extra drift of $\frac{1}{n} \mu(Y_{\frac{i}{n}}^{(n)})$. If we then interpolate *this* function to a function $\{\overline{Y}_t^{(n)}\}$ defined for all $t \geq 0$, and let $n \to \infty$, we obtain a *diffusion*, with *drift* $\mu(x)$, and *diffusion coefficient* or *volatility* $\sigma(x)$. Intuitively, the diffusion is like Brownian motion, except that its mean is drifting according to the function $\mu(x)$, and its local variability is scaled according to the function $\sigma(x)$. In this context, Brownian motion is the special case $\mu(x) \equiv 0$ and $\sigma(x) \equiv 1$.

Now, for large n we see from (15.4.1) that $\frac{1}{\sqrt{n}} Z_{i+1} \approx B_{\frac{i+1}{n}} - B_{\frac{i}{n}}$, so that (15.6.1) is essentially saying that

$$Y_{\frac{i+1}{n}}^{(n)} - Y_{\frac{i}{n}}^{(n)} \approx \left(B_{\frac{i+1}{n}} - B_{\frac{i}{n}} \right) \sigma \left(Y_{\frac{i}{n}}^{(n)} \right) + \frac{1}{n} \mu \left(Y_{\frac{i}{n}}^{(n)} \right) .$$

If we now (intuitively) set $X_t = \lim_{n \to \infty} \overline{Y}_t^{(n)}$ for each fixed $t \geq 0$, then we can write the limiting version of (15.6.1) as

$$dX_t = \sigma(X_t)\, dB_t + \mu(X_t)\, dt , \qquad (15.6.2)$$

where $\{B_t\}_{t \geq 0}$ is Brownian motion. The process $\{X_t\}_{t \geq 0}$ is called a *diffusion* with drift $\mu(x)$ and diffusion coefficient or *volatility* $\sigma(x)$. Its definition (15.6.2) can be interpreted roughly as saying that, as $h \searrow 0$, we have

$$X_{t+h} \approx X_t + \sigma(X_t)(B_{t+h} - B_t) + \mu(X_t)\, h ; \qquad (15.6.3)$$

thus, given that $X_t = x$, we have that approximately $X_{t+h} \sim N(x + \mu(x)\, h,\ \sigma^2(x)\, h)$. If $\mu(x) \equiv 0$ and $\sigma(x) \equiv 1$, then the diffusion coincides exactly with Brownian motion.

We can also compute a *generator* for a diffusion defined by (15.6.2). To do this we must generalise the matrix generator (15.3.4) for processes on finite state spaces. We define the generator to be an operator $\mathcal{Q}$ acting on smooth functions $f : \mathbf{R} \to \mathbf{R}$ by

$$(\mathcal{Q}f)(x) = \lim_{h \to 0} \frac{1}{h} \mathbf{E}_x \left(f(X_{t+h}) - f(X_t) \right) , \qquad (15.6.4)$$

where $\mathbf{E}_x$ means expectation conditional on $X_t = x$. That is, $\mathcal{Q}$ maps one smooth function f to another function, $\mathcal{Q}f$.

Now, for a diffusion satisfying (15.6.3), we can use a Taylor series expansion to approximate $f(X_{t+h})$ as

$$f(X_{t+h}) \approx f(X_t) + (X_{t+h}-X_t)f'(X_t) + \frac{1}{2}(X_{t+h}-X_t)^2\, f''(X_t)\,.\quad (15.6.5)$$

On the other hand,

$$X_{t+h} - X_t \approx \sigma(X_t)(B_{t+h} - B_t) + \mu(X_t)h\,.$$

Conditional on $X_t = x$, the conditional distribution of $B_{t+h} - B_t$ is normal with mean 0 and variance h; hence, the conditional distribution of $X_{t+h}-X_t$ is approximately normal with mean $\mu(x)\,h$ and variance $\sigma^2(x)\,h$. So, taking expected values conditional on $X_t = x$, we conclude from (15.6.5) that

$$\mathbf{E}_x\left[f(X_{t+h}) - f(X_t)\right] \approx \mu(x)hf'(x) + \frac{1}{2}\left[(\mu(x)h)^2 + \sigma^2(x)h\right]\, f''(x)$$

$$= \mu(x)hf'(x) + \frac{1}{2}\sigma^2(x)hf''(x) + O(h^2)\,.$$

It then follows from the definition (15.6.4) of generator that

$$(\mathcal{Q}f)(x) = \mu(x)\,f'(x) + \frac{1}{2}\,\sigma^2(x)\,f''(x)\,.\quad (15.6.6)$$

We have thus given an explicit formula for the generator $\mathcal{Q}$ of a diffusion defined by (15.6.2). Indeed, as with Markov processes on discrete spaces, diffusions are often characterised in terms of their generators, which again provide all the information necessary to completely reconstruct the process's transition probabilities.

Finally, we note that Brownian motion can be used to define a certain unusual kind of integral, called a *stochastic integral* or *Itô integral*. Let $\{B_t\}_{t\geq 0}$ be Brownian motion, and let $\{C_t\}_{t\geq 0}$ be a *non-anticipative* real-valued random variable (i.e., the value of C_t is conditionally independent of $\{B_s\}_{s>t}$, given $\{B_s\}_{s\leq t}$). Then it is possible to define an integral of the form $\int_a^b C_t\, dB_t$, i.e. an integral "with respect to Brownian motion". Roughly speaking, this integral is defined by

$$\int_a^b C_t\, dB_t = \lim \sum_{i=0}^{m-1} C_{t_i}(B_{t_{i+1}} - B_{t_i})\,,$$

where $a = t_0 \leq t_1 \leq \ldots \leq t_m = b$, and where the limit is over finer and finer partitions $\{(t_{i-1}, t_i]\}$ of (a, b). Note that this integral is random both because of the randomness of the values of the integrand C_t, and also because of the randomness from Brownian motion B_t itself. In particular, we now see that (15.6.2) can be written equivalently as

$$X_b - X_a = \int_a^b \sigma(X_t) \, dB_t + \int_a^b \mu(X_t) \, dt,$$

where the first integral is with respect to Brownian motion. Such stochastic integrals are used in a variety of research areas; for further details see e.g. Bhattacharya and Waymire (1990, Chapter VII), or the other references in Subsection B.5.

Exercise 15.6.7. Let $\{X_t\}_{t \geq 0}$ be a diffusion with constant drift $\mu(x) \equiv a$ and constant volatility $\sigma(x) \equiv b > 0$.
(a) Show that $X_t = a\,t + b\,B_t$.
(b) Show that $\mathcal{L}(X_t) = N(at, \, b^2 t)$.
(c) Compute $\mathbf{E}(X_s\, X_t)$ for $0 < s < t$.

Exercise 15.6.8. Let $\{B_t\}_{t \geq 0}$ be standard Brownian motion, with $B_0 = 0$. Let $X_t = \int_0^t s \, ds + \int_0^t b\, B_s \, ds = at + bB_t$ be a diffusion with constant drift $\mu(x) \equiv a > 0$ and constant volatility $\sigma(x) \equiv b > 0$. Let $Z_t = \exp\left[-2aX_t/b^2\right]$.
(a) Prove that $\{Z_t\}_{t \geq 0}$ is a martingale, i.e. that $\mathbf{E}[Z_t \mid Z_u \, (0 \leq u \leq s)] = Z_s$ for $0 < s < t$.
(b) Let $A, B > 0$, and let $T_A = \inf\{t \geq 0 : X_t = A\}$ and $T_{-B} = \inf\{t \geq 0 : X_t = -B\}$ denote the first hitting times of A and $-B$, respectively. Compute $P(T_A < T_{-B})$. [Hint: Use part (a); you may assume that Corollary 14.1.7 also applies in continuous time.]

Exercise 15.6.9. Let $g : \mathbf{R} \to \mathbf{R}$ be a smooth function with $g(x) > 0$ and $\int g(x)dx = 1$. Consider the diffusion $\{X_t\}$ defined by (15.6.2) with $\sigma(x) \equiv 1$ and $\mu(x) = \frac{1}{2}g'(x)/g(x)$. Show that the probability distribution having density g (with respect to Lebesgue measure on $\mathbf{R}$) is a stationary distribution for the diffusion. [Hint: Show that the diffusion is *reversible* with respect to $g(x)\,dx$. You may use (15.6.3) and Exercise 15.3.7.] This diffusion is called a *Langevin diffusion*.

Exercise 15.6.10. Let $\{p_{ij}^t\}$ be the transition probabilities for a Markov chain on a finite state space $\mathcal{X}$. Define the matrix $Q = (q_{ij})$ by (15.3.4). Let $f : \mathcal{X} \to \mathbf{R}$ be any function, and let $i \in \mathcal{X}$. Prove that $(Qf)_i$ (i.e., $\sum_{k \in \mathcal{X}} q_{ik} f_k$) corresponds to $(\mathcal{Q}f)(i)$ from (15.6.4).

Exercise 15.6.11. Suppose a diffusion $\{X_t\}_{t \geq 0}$ satisfies that

$$(\mathcal{Q}f)(x) = 17 f'(x) + 23\, x^2 \, f''(x),$$

for all smooth $f : \mathbf{R} \to \mathbf{R}$. Compute the drift and volatility functions for this diffusion.

Exercise 15.6.12.   Suppose a diffusion $\{X_t\}_{t\geq 0}$ has generator given by $(\mathcal{Q}f)(x) = -\frac{x}{2}f'(x) + \frac{1}{2}f''(x)$. (Such a diffusion is called an *Ornstein-Uhlenbeck process.*)
(a) Write down a formula for dX_t. [Hint: Use (15.6.6).]
(b) Show that $\{X_t\}_{t\geq 0}$ is reversible with respect to the standard normal distribution, $N(0, 1)$. [Hint: Use Exercise 15.6.9.]

15.7. Itô's Lemma.

Let $\{X_t\}_{t\geq 0}$ be a diffusion satisfying (15.6.2), and let $f : \mathbf{R} \to \mathbf{R}$ be a smooth function. Then we might expect that $\{f(X_t)\}_{t\geq 0}$ would itself be a diffusion, and we might wonder what drift and volatility would correspond to $\{f(X_t)\}_{t\geq 0}$. Specifically, we would like to compute an expression for $d(f(X_t))$.

To that end, we consider small $h \approx 0$, and use the Taylor approximation

$$f(X_{t+h}) \approx f(X_t) + f'(X_t)(X_{t+h} - X_t) + \frac{1}{2}f''(X_t)(X_{t+h} - X_t)^2 + o(h).$$

Now, in the classical world, an expression like $(X_{t+h} - X_t)^2$ would be $O(h^2)$ and hence could be neglected in this computation. However, we shall see that for Itô integrals this is not the case.

We continue by writing

$$X_{t+h} - X_t = \int_t^{t+h} \sigma(X_s)dB_s + \int_t^{t+h} \mu(X_s)ds$$

$$\approx \sigma(X_t)(B_{t+h} - B_t) + \mu(X_t)h.$$

Hence,

$$f(X_{t+h}) \approx f(X_t) + f'(X_t)[\sigma(X_t)(B_{t+h} - B_t) + \mu(X_t)h]$$

$$+ \frac{1}{2}f''(X_t)[\sigma(X_t)(B_{t+h} - B_t) + \mu(X_t)h]^2 + o(h).$$

In the expansion of $(\sigma(X_t)(B_{t+h} - B_t) + \mu(X_t)h)^2$, all terms are $o(h)$ aside from the first, so that

$$f(X_{t+h}) \approx f(X_t) + f'(X_t)[\sigma(X_t)(B_{t+h} - B_t) + \mu(X_t)h]$$

$$+ \frac{1}{2}f''(X_t)[\sigma(X_t)(B_{t+h} - B_t)]^2 + o(h).$$

But $B_{t+h} - B_t \sim N(0, h)$, so that $\mathbf{E}\left[(B_{t+h} - B_t)^2\right] = h$, and in fact $(B_{t+h} - B_t)^2 = h + o(h)$. (This is the departure from classical integration theory; there, we would have $(B_{t+h} - B_t)^2 = o(h)$.) Hence, as $h \to 0$,

$$f(X_{t+h}) \approx f(X_t) + f'(X_t)\left[\sigma(X_t)(B_{t+h} - B_t) + \mu(X_t) h\right]$$

$$+ \frac{1}{2} f''(X_t) \sigma^2(X_t) h + o(h).$$

We can write this in differential form as

$$d\left(f(X_t)\right) = f'(X_t)\left[\sigma(X_t) dB_t + \mu(X_t) dt\right] + \frac{1}{2} f''(X_t) \sigma^2(X_t) dt,$$

or

$$d\left(f(X_t)\right) = f'(X_t) \sigma(X_t) dB_t + \left(f'(X_t) \mu(X_t) + \frac{1}{2} f''(X_t) \sigma^2(X_t)\right) dt.$$
$$(15.7.1)$$

That is, $\{f(X_t)\}$ is itself a diffusion, with new drift $\overline{\mu}(x) = f'(x) \mu(x) + \frac{1}{2} f''(x) \sigma^2(x)$ and new volatility $\overline{\sigma}(x) = f'(x) \sigma(x)$.

Equation (15.7.1) is *Itô's Lemma*, or *Itô's formula*. It is a generalisation of the classical chain rule, and is very important for doing computations with Itô integrals. By (15.6.2), it implies that $d(f(X_t))$ is equal to $f'(X_t) dX_t + \frac{1}{2} f''(X_t) \sigma^2(X_t) dt$. The extra term $\frac{1}{2} f''(X_t) \sigma^2(X_t) dt$ arises because of the unusual nature of Brownian motion, namely that $(B_{t+h} - B_t)^2 \approx h$ instead of being $O(h^2)$. (In particular, this again indicates why Brownian motion is not differentiable; compare Exercise 15.4.6.)

Exercise 15.7.2. Consider the Ornstein-Uhlenbeck process $\{X_t\}$ of Exercise 15.6.12, with generator $(\mathcal{Q}f)(x) = -\frac{x}{2} f'(x) + \frac{1}{2} f''(x)$.
(a) Let $Y_t = (X_t)^2$ for each $t \geq 0$. Compute dY_t.
(b) Let $Z_t = (X_t)^3$ for each $t \geq 0$. Compute dZ_t.

Exercise 15.7.3. Consider the diffusion $\{X_t\}$ of Exercise 15.6.11, with generator $(\mathcal{Q}f)(x) = 17 f'(X_t) + 23(X_t)^2 f''(X_t)$. Let $W_t = (X_t)^5$ for each $t \geq 0$. Compute dW_t.

15.8. The Black-Scholes equation.

In mathematical finance, the prices of stocks are often modeled as diffusions. Various calculations, such as the value of complicated *stock options* or *derivatives*, then proceed on that basis.

In this final subsection, we give a very brief introduction to financial models, and indicate the derivation of the famous *Black-Scholes equation* for pricing a certain kind of stock option. Our treatment is very cursory; for more complete discussion see books on mathematical finance, such as those listed in Subsection B.6.

For simplicity, we suppose there is a constant "risk-free" interest rate r. This means that an amount of money M at present is worth $e^{rt}M$ at a time t later. Equivalently, an amount of money M a time t later is worth only $e^{-rt}M$ at present.

We further suppose that there is a "stock" available for purchase. This stock has a purchase price P_t at time t. It is assumed that this stock price can be modeled as a diffusion process, according to the equation

$$dP_t = bP_t dt + \sigma P_t dB_t, \qquad (15.8.1)$$

i.e. with $\mu(x) = bx$ and $\sigma(x) = \sigma x$. Here B_t is Brownian motion, b is the *appreciation rate*, and σ is the *volatility* of the stock. For present purposes we assume that b and σ are constant, though more generally they (and also r) could be functions of t.

We wish to find a value, or fair price, for a "European call option" (an example of a financial *derivative*). A European call option is the option to purchase the stock at a fixed time $T > 0$ for a fixed price $q > 0$. Obviously, one would exercise this option if and only if $P_T > q$, and in this case one would gain an amount $P_T - q$. Therefore, the value at time T of the European call option will be precisely $\max(0, P_T - q)$; it follows that the value at time 0 is $e^{-rT}\max(0, P_T - q)$. The problem is that, at time 0, the future price P_T is unknown (i.e. random). Hence, at time 0, the true value of the option is also unknown. What we wish to compute is a fair price for this option at time 0, given only the current price P_0 of the stock.

To specify what a fair price means is somewhat subtle. This price is taken to be the (unique) value such that neither the buyer nor the seller could strictly improve their payoff by investing in the stock directly (as opposed to investing in the option), no matter how sophisticated an investment strategy they use (including selling short, i.e. holding a *negative* number of shares) and no matter how often they buy and sell the stock (without being charged any transaction commissions).

It turns out that for our model, the fair price of the option is equal to the *expected value* of the option given P_0, but only after replacing b by r, i.e. after replacing the appreciation rate by the risk-free interest rate. We do not justify this fact here; for details see e.g. Theorem 1.2.1 of Karatzas, 1997. The argument involves switching to a different probability measure (the *risk-neutral equivalent martingale measure*), under which $\{B_t\}$ is still a (different) Brownian motion, but where (15.8.1) is replaced by $dP_t = rP_t dt + \sigma P_t dB_t$.

We are thus interested in computing

$$\mathbf{E}\left(e^{-rT} \max(0, \ P_T - q) \,|\, P_0\right),\tag{15.8.2}$$

under the condition that

$$b \text{ is replaced by } r.\tag{15.8.3}$$

To proceed, we use Itô's Lemma to compute $d(\log P_t)$. This means that we let $f(x) = \log x$ in (15.7.1), so that $f'(x) = 1/x$ and $f''(x) = -1/x^2$. We apply this to the diffusion (15.8.1), under the condition (15.8.3), so that $\mu(x) = rx$ and $\sigma(x) = \sigma x$. We compute that

$$
\begin{aligned}
d(\log P_t) &= \frac{1}{P_t}\sigma P_t\, dB_t + \left(rP_t\frac{1}{P_t} + \frac{1}{2}\left(-\frac{1}{P_t^2}\right)(\sigma P_t)^2\right) dt \\
&= \sigma\, dB_t + \left(r - \frac{1}{2}\sigma^2\right) dt.
\end{aligned}
$$

Since these coefficients are all constants, it now follows that

$$\log(P_T/P_0) = \log P_T - \log P_0 = \sigma(B_T - B_0) + \left(r - \frac{1}{2}\sigma^2\right)(T - 0),$$

whence

$$P_T = P_0 \exp\left(\sigma(B_T - B_0) + \left(r - \frac{1}{2}\sigma^2\right)T\right).$$

Recall now that $B_T - B_0$ has the normal distribution $N(0, T)$, with density function $\frac{1}{\sqrt{2\pi T}}e^{-x^2/2T}$. Hence, by Proposition 6.2.3, we can compute the value (15.8.2) of the European call option as

$$\int_{-\infty}^{\infty} e^{-rT} \max\left(0, \ P_0 \exp\left(\sigma x + (r - \frac{1}{2}\sigma^2)T\right) - q\right) \frac{1}{\sqrt{2\pi T}} e^{-x^2/2T}\, dx.$$
$$\tag{15.8.4}$$

After some re-arranging and substituting, we simplify this integral to

$$P_0\, \Phi\left(\frac{1}{\sigma\sqrt{T}}\left(\log(P_0/q) + T(r + \frac{1}{2}\sigma^2)\right)\right)$$

$$- qe^{-rT}\Phi\left(\frac{1}{\sigma\sqrt{T}}\left(\log(P_0/q) + T(r - \frac{1}{2}\sigma^2)\right)\right),\tag{15.8.5}$$

where $\Phi(z) = \frac{1}{\sqrt{2\pi}}\int_{-\infty}^{z} e^{-s^2/2}ds$ is the usual cumulative distribution function of the standard normal.

Equation (15.8.5) thus gives the time-0 value, or fair price, of a European call option to buy at time $T > 0$ for price $q > 0$ a stock having initial price

P_0 and price process $\{P_t\}_{t\geq 0}$ governed by (15.8.1). Equation (15.8.5) is the well-known *Black-Scholes equation* for computing the price of a European call option. Recall that here r is the risk-free interest rate, and σ is the volatility. Furthermore, the appreciation rate b does not appear in the final formula, which is advantageous because b is difficult to estimate in practice.

For further details of this subject, including generalisations to more complicated financial models, see any introductory book on the subject of mathematical finance, such as those listed in Subsection B.6.

Exercise 15.8.6.   Show that (15.8.4) is indeed equal to (15.8.5).

Exercise 15.8.7.   Consider the price formula (15.8.5), with r, σ, T, and P_0 fixed positive quantities.
(a) What happens to the price (15.8.5) as $q \searrow 0$? Does this result make intuitive sense?
(b) What happens to the price (15.8.5) as $q \to \infty$? Does this result make intuitive sense?

Exercise 15.8.8.   Consider the price formula (15.8.5), with r, σ, P_0, and q fixed positive quantities.
(a) What happens to the price (15.8.5) as $T \searrow 0$? [Hint: Consider separately the cases $q > P_0$, $q = P_0$, and $q < P_0$.] Does this result make intuitive sense?
(b) What happens to the price (15.8.5) as $T \to \infty$? Does this result make intuitive sense?

15.9. Section summary.

This section considered various aspects of the theory of stochastic processes, in a somewhat informal manner. All of the topics considered can be thought of as generalisations of the discrete-time, discrete-space processes of Sections 7 and 8.

First, the section considered existence questions, and stated (but did not prove) the important Kolmogorov Existence Theorem. It then discussed discrete-time processes on general (non-discrete) state spaces. It provided generalised notions of irreducibility and aperiodicity, and stated a theorem about convergence of irreducible, aperiodic Markov chains to their stationary distributions. Next, continuous-time processes were considered. The semigroup property of Markov transition probabilities was presented. Generators of continuous-time, discrete-space Markov processes were discussed.

The section then focused on processes in continuous time and space. Brownian motion was developed as a limit of random walks that take smaller and smaller steps, more and more frequently; the normal, independent increments of Brownian motion then followed naturally. Diffusions were then developed as generalisations of Brownian motion. Generators of diffusions were described, as were stochastic integrals.

The section closed with intuitive derivations of Itô's Lemma, and of the Black-Scholes equation from mathematical finance. The reader was encouraged to consult the books of Subsections B.5 and B.6 for further details.

A. Mathematical Background.

This section reviews some of the mathematics that is necessary as a prerequisite to understanding this text. The material is reviewed in a form which is most helpful for our purposes. Note, however, that this material is merely an *outline* of what is needed; if this section is not largely review for you, then perhaps your background is insufficient to follow this text in detail, and you may wish to first consult references such as those in Subsection B.1.

A.1. Sets and functions.

A fundamental object in mathematics is a *set*, which is any[*] collection of objects. For example, $\{1, 3, 4, 7\}$, the collection of rational numbers, the collection of real numbers, and the empty set $\emptyset$ containing no elements, are all sets. Given a set, a *subset* of that set is any subcollection. For example, $\{3, 7\}$ is a subset of $\{1, 3, 4, 7\}$; in symbols, $\{3, 7\} \subseteq \{1, 3, 4, 7\}$.

Given a subset, its *complement* is the set consisting of all elements of a "universal" set which are *not* in the subset. In symbols, if $A \subseteq \Omega$, where Ω is understood to be the universal set, then $A^C = \{\omega \in \Omega; \omega \notin A\}$.

Given a collection of sets $\{A_\alpha\}_{\alpha \in I}$ (where I is any indicator set; perhaps $I = \{1, 2\}$), we define their *union* to be the set of all elements which are in *at least one* of the A_α; in symbols, $\bigcup_\alpha A_\alpha = \{\omega; \omega \in A_\alpha \text{ for some } \alpha \in I\}$. Similarly, we define their *intersection* to be the set of all elements which are in *all* of the A_α; in symbols, $\bigcap_\alpha A_\alpha = \{\omega; \omega \in A_\alpha \text{ for all } \alpha \in I\}$.

Union, intersection, and complement satisfy *de Morgan's Laws*, which state that $\left(\bigcup_\alpha A_\alpha\right)^C = \bigcap_\alpha A_\alpha^C$ and $\left(\bigcap_\alpha A_\alpha\right)^C = \bigcup_\alpha A_\alpha^C$. In words, complement converts unions to intersections and vice-versa.

A collection $\{A_\alpha\}_{\alpha \in I}$ of subsets of a set Ω are *disjoint* if the intersection of any pair of them is the empty set; in symbols, if $A_{\alpha_1} \cap A_{\alpha_2} = \emptyset$ whenever $\alpha_1, \alpha_2 \in I$ are distinct. The collection $\{A_\alpha\}_{\alpha \in I}$ is called a *partition* of Ω if $\{A_\alpha\}_{\alpha \in I}$ is disjoint and also $\bigcup_\alpha A_\alpha = \Omega$. If $\{A_\alpha\}_{\alpha \in I}$ are disjoint, we sometimes write their union as $\overset{\bullet}{\underset{\alpha \in I}{\bigcup}} A_\alpha$ (e.g. $A_{\alpha_1} \overset{\bullet}{\cup} A_{\alpha_2}$) and call it a *disjoint union*.

We also define the *difference* of sets A and B by $A \setminus B = A \cap B^C$; for example, $\{1, 3, 4, 7\} \setminus \{3, 7, 9\} = \{1, 4\}$.

Given sets Ω_1 and Ω_2, we define their *(Cartesian) product* to be the set of all ordered pairs having first element from Ω_1 and second element from

---

[*]In fact, certain very technical restrictions are required to avoid contradictions such as Russell's paradox. But we do not consider that here.

Ω_2; in symbols, $\Omega_1 \times \Omega_2 = \{(\omega_1, \omega_2); \omega_1 \in \Omega_1, \ \omega_2 \in \Omega_2\}$. Similarly, larger Cartesian products $\prod_\alpha \Omega_\alpha$ are defined.

A *function* f is a mapping* from a set Ω_1 to a second set Ω_2; in symbols, $f : \Omega_1 \to \Omega_2$. Its *domain* is the set of points in Ω_1 for which it is defined, and its *range* is the set of points in Ω_2 which are of the form $f(\omega_1)$ for some $\omega_1 \in \Omega_1$.

If $f : \Omega_1 \to \Omega_2$, and if $S_1 \subseteq \Omega_1$ and $S_2 \subseteq \Omega_2$, then the *image* of S_1 under f is given by $f(S_1) = \{\omega_2 \in \Omega_2; \omega_2 = f(\omega_1)$ for some $\omega_1 \in S_1\}$, and the *inverse image* of S_2 under f is given by $f^{-1}(S_2) = \{\omega_1 \in \Omega_1; f(\omega_1) \in S_2\}$. We note that inverse images preserve the usual set operations, for example $f^{-1}(S_1 \cup S_2) = f^{-1}(S_1) \cup f^{-1}(S_2); f^{-1}(S_1 \cap S_2) = f^{-1}(S_1) \cap f^{-1}(S_2);$ and $f^{-1}(S^C) = f^{-1}(S)^C$.

Special sets include the integers $\mathbf{Z} = \{0, 1, -1, 2, -2, 3, \ldots\}$, the positive integers or natural numbers $\mathbf{N} = \{1, 2, 3, \ldots\}$, the rational numbers $\mathbf{Q} = \{\frac{m}{n}; m, n \in \mathbf{Z}, n \neq 0\}$, and the real numbers $\mathbf{R}$.

Given a subset $S \subseteq \Omega$, its *indicator function* $\mathbf{1}_S : \Omega \to \mathbf{R}$ is defined by

$$\mathbf{1}_S(\omega) = \begin{cases} 1, & \omega \in S \\ 0, & \omega \notin S. \end{cases}$$

Finally, the *Axiom of Choice* states that given a collection $\{A_\alpha\}_{\alpha \in I}$ of non-empty sets (i.e., $A_\alpha \neq \emptyset$ for all α), their Cartesian product $\prod_\alpha A_\alpha$ is also non-empty. In symbols,

$$\prod_\alpha A_\alpha \neq \emptyset \qquad \text{whenever} \qquad A_\alpha \neq \emptyset \text{ for each } \alpha. \qquad (A.1.1)$$

That is, it is possible to "choose" one element simultaneously from each of the non-empty sets A_α – a fact that seems straightforward, but does not follow from the other axioms of set theory.

A.2. Countable sets.

One important property of sets is their *cardinality*, or size. A set Ω is *finite* if for some $n \in \mathbf{N}$ and some function $f : \mathbf{N} \to \Omega$, we have $f(\{1, 2, \ldots, n\}) \supseteq \Omega$. A set is *countable* if for some function $f : \mathbf{N} \to \Omega$, we have $f(\mathbf{N}) = \Omega$. (Note that, by this definition, all finite sets are countable. As a special case, the empty set $\emptyset$ is both finite and countable.) A set is *countably infinite* if it is countable but not finite. It is *uncountable* if it is not countable.

---

*More formally, it is a collection of ordered pairs $(\omega_1, \omega_2) \in \Omega_1 \times \Omega_2$, with $\omega_1 \in \Omega_1$ and $\omega_2 = f(\omega_1) \in \Omega_2$, such that no ω_1 appears in more than one pair.

Intuitively, a set is countable if its elements can be arranged in a *sequence*. It turns out that countability of sets is very important in probability theory. The axioms of probability theory make certain guarantees about countable operations (e.g., the countable additivity of probability measures) which do *not* necessarily hold for uncountable operations. Thus, it is important to understand which sets are countable and which are not.

The natural numbers **N** are obviously countable; indeed, we can take the identity function (i.e., $f(n) = n$ for all $n \in \mathbf{N}$) in the definition of countable above. Furthermore, it is clear that any subset of a countable set is again countable; hence, any subset of **N** is also countable.

The integers **Z** are also countable: we can take, say, $f(1) = 0$, $f(2) = 1$, $f(3) = -1$, $f(4) = 2$, $f(5) = -2$, $f(6) = 3$, etc. to ensure that $f(\mathbf{N}) = \mathbf{Z}$. More surprising is

Proposition A.2.1. *Let Ω_1 and Ω_2 be countable sets. Then their Cartesian product $\Omega_1 \times \Omega_2$ is also countable.*

Proof. Let f_1 and f_2 be such that $f_1(\mathbf{N}) = \Omega_1$ and $f_2(\mathbf{N}) = \Omega_2$. Then define a new function $f : \mathbf{N} \to \Omega_1 \times \Omega_2$ by $f(1) = (f_1(1), f_2(1))$, $f(2) = (f_1(1), f_2(2))$, $f(3) = (f_1(2), f_2(1))$, $f(4) = (f_1(1), f_2(3))$, $f(5) = (f_1(2), f_2(2))$, $f(6) = (f_1(3), f_2(1))$, $f(7) = (f_1(1), f_2(4))$, etc. (See Figure A.2.2.) Then $f(\mathbf{N}) = \Omega_1 \times \Omega_2$, as required. ∎

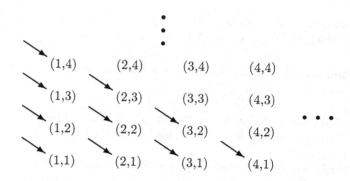

Figure A.2.2. Constructing f in Proposition A.2.1.

From this proposition, we see that e.g. $\mathbf{N} \times \mathbf{N}$ is countable, as is $\mathbf{Z} \times \mathbf{Z}$. It follows immediately (and, perhaps, surprisingly) that the set of all rational numbers is countable; indeed, **Q** is equivalent to a subset of $\mathbf{Z} \times \mathbf{Z}$

if we identify the rational number m/n (in lowest terms) with the element $(m, n) \in \mathbf{Z} \times \mathbf{Z}$. It also follows that, given a sequence $\Omega_1, \Omega_2, \ldots$ of countable sets, their countable union $\Omega = \bigcup_i \Omega_i$ is also countable; indeed, if $f_i(\mathbf{N}) = \Omega_i$, then we can identify Ω with a subset of $\mathbf{N} \times \mathbf{N}$ by the mapping $(m, n) \mapsto f_m(n)$.

For another example of the use of Proposition A.2.1, recall that a real number is *algebraic* if it is a root of some non-constant polynomial with integer coefficients.

Exercise A.2.3. Prove that the set of all algebraic numbers is countable.

On the other hand, the set of all *real* numbers is *not* countable. Indeed, even the unit interval $[0, 1]$ is not countable. To see this, suppose to the contrary that it were countable, with $f : \mathbf{N} \to [0, 1]$ such that $f(\mathbf{N}) = [0, 1]$. Imagine writing each element of $[0, 1]$ in its usual base-10 expansion, and let d_i be the ith digit of the number $f(i)$. Now define c_i by: $c_i = 2$ if $d_i \geq 5$, while $c_i = 7$ if $d_i < 5$. (In particular, $c_i \neq d_i$.) Then let $x = \sum_{i=1}^{\infty} c_i 10^{-i}$ (so that the base-10 expansion of x is $0.c_1 c_2 c_3 \ldots$). Then x differs from $f(i)$ in the ith digit of their base-10 expansions, for each i. Hence, $x \neq f(i)$ for any i. This contradicts the assumption that $f(\mathbf{N}) = [0, 1]$.

A.3. Epsilons and Limits.

Real analysis and probability theory make constant use of arguments involving arbitrarily small $\epsilon > 0$. A basic starting block is

Proposition A.3.1. *Let a and b be two real numbers. Suppose that, for any $\epsilon > 0$, we have $a \leq b + \epsilon$. Then $a \leq b$.*

Proof. Suppose to the contrary that $a > b$. Let $\epsilon = \frac{a-b}{2}$. Then $\epsilon > 0$, but it is not the case that $a \leq b + \epsilon$. This is a contradiction. ∎

Arbitrarily small $\epsilon > 0$ are used to define the important concept of a *limit*. We say that a sequence of real numbers $x_1, x_2, \ldots$ *converges* to the real number x (or, has *limit* x) if, given any $\epsilon > 0$, there is $N \in \mathbf{N}$ (which may depend on ϵ) such that for any $n \geq N$, we have $|x_n - x| < \epsilon$. We shall also write this as $\lim_{n \to \infty} x_n = x$, and shall sometimes abbreviate this to $\lim_n x_n = x$ or even $\lim x_n = x$. We also write this as $\{x_n\} \to x$.

Exercise A.3.2. Use the definition of a limit to prove that
(a) $\lim_{n \to \infty} \frac{1}{n} = 0$.

(b) $\lim_{n\to\infty} \frac{1}{n^k} = 0$ for any $k > 0$.

(c) $\lim_{n\to\infty} 2^{1/n} = 1$.

A useful reformulation of limits is given by

Proposition A.3.3. *Let $\{x_n\}$ be a sequence of real numbers. Then $\{x_n\} \to x$ if and only if for each $\epsilon > 0$, the set $\{n \in \mathbf{N};\ |x_n - x| \geq \epsilon\}$ is finite.*

Proof. If $\{x_n\} \to x$, then given $\epsilon > 0$, there is $N \in \mathbf{N}$ such that $|x_n - x| < \epsilon$ for all $n \geq N$. Hence, the set $\{n;\ |x_n - x| \geq \epsilon\}$ contains at most $N - 1$ elements, and so is finite.

Conversely, given $\epsilon > 0$, if the set $\{n;\ |x_n - x| \geq \epsilon\}$ is finite, then it has some largest element, say K. Setting $N = K + 1$, we see that $|x_n - x| < \epsilon$ whenever $n \geq N$. Hence, $\{x_n\} \to x$. ∎

Exercise A.3.4. Use Proposition A.3.3 to provide an alternative proof for each of the three parts of Exercise A.3.2.

A sequence which does not converge is said to *diverge*. There is one special case. A sequence $\{x_n\}$ *converges to infinity* if for all $M \in \mathbf{R}$, there is $N \in \mathbf{N}$, such that $x_n \geq M$ whenever $n \geq N$. (Similarly, $\{x_n\}$ converges to negative infinity if for all $M \in \mathbf{R}$, there is $N \in \mathbf{N}$, such that $x_n \leq M$ whenever $n \geq N$.) This is a special kind of "convergence" which is also sometimes referred to as divergence!

Limits have many useful properties. For example,

Exercise A.3.5. Prove the following:
(a) If $\lim_n x_n = x$, and $a \in \mathbf{R}$, then $\lim_n a x_n = ax$.
(b) If $\lim_n x_n = x$ and $\lim_n y_n = y$, then $\lim_n (x_n + y_n) = x + y$.
(c) If $\lim_n x_n = x$ and $\lim_n y_n = y$, then $\lim_n (x_n y_n) = xy$.
(d) If $\lim_n x_n = x$, and $x \neq 0$, then $\lim_n (1/x_n) = 1/x$.

Another useful property is given by

Proposition A.3.6. *Suppose $x_n \to x$, and $x_n \leq a$ for all $n \in \mathbf{N}$. Then $x \leq a$.*

Proof. Suppose to the contrary that $x > a$. Let $\epsilon = \frac{x-a}{2}$. Then $\epsilon > 0$, but for all $n \in \mathbf{N}$ we have $|x_n - x| \geq x - x_n \geq x - a > \epsilon$, contradicting the assertion that $x_n \to x$. ∎

Given a sequence $\{x_n\}$, a *subsequence* $\{x_{n_k}\}$ is any sequence formed from $\{x_n\}$ by omitting some of the elements. (For example, one subsequence of the sequence $(1, 3, 5, 7, 9, \ldots)$ of all positive odd numbers is the sequence $(3, 9, 27, 81, \ldots)$ of all positive integer powers of 3.) The *Bolzano-Weierstrass theorem*, a consequence of compactness, says that every bounded sequence contains a convergent subsequence.

Limits are also used to define infinite series. Indeed, the sum $\sum_{i=1}^{\infty} s_i$ is simply a shorthand way of writing $\lim_{n \to \infty} \sum_{i=1}^{n} s_i$. If this limit exists and is finite, then we say that the series $\sum_{i=1}^{\infty} s_i$ *converges*; otherwise we say the series *diverges*. (In particular, for infinite series, converging to infinity is usually referred to as diverging.) If the s_i are non-negative, then $\sum_{i=1}^{\infty} s_i$ either converges to a finite value or diverges to infinity; we write this as $\sum_{i=1}^{\infty} s_i < \infty$ and $\sum_{i=1}^{\infty} s_i = \infty$, respectively.

For example, it is not hard to show (by comparing $\sum_{i=1}^{\infty} i^{-a}$ to the integral $\int_1^{\infty} t^{-a} dt$) that

$$\sum_{i=1}^{\infty} (1/i^a) < \infty \qquad \text{if and only if} \qquad a > 1. \qquad (A.3.7)$$

Exercise A.3.8.   Prove that $\sum_{i=2}^{\infty} \left(1 / i \log(i) \right) = \infty$. [Hint: Show $\sum_{i=2}^{\infty} (1 / i \log(i)) \geq \int_2^{\infty} (dx / x \log x).$]

Exercise A.3.9.   Prove that $\sum_{i=3}^{\infty} \left(1 / i \log(i) \log \log(i) \right) = \infty$, but $\sum_{i=3}^{\infty} \left(1 / i \log(i) \left[\log \log(i) \right]^2 \right) < \infty$.

A.4. Infimums and supremums.

Given a set $\{x_\alpha\}_{\alpha \in I}$ of real numbers, a *lower bound* for them is a real number ℓ such that $x_\alpha \geq \ell$ for all $\alpha \in I$. The set $\{x_\alpha\}_{\alpha \in I}$ is *bounded below* if it has at least one lower bound. A lower bound is called the *greatest lower bound* if it is at least as large as any other lower bound.

A very important property of the real numbers is: *Any non-empty set of real numbers which is bounded below has a unique greatest lower bound.* The uniqueness part of this is rather obvious; however, the existence part is very subtle. For example, this property would *not* hold if we restricted ourselves to rational numbers. Indeed, the set $\{q \in \mathbf{Q}; q > 0, q^2 > 2\}$ does not have a greatest lower bound if we allow rational numbers only; however, if we allow all real numbers, then the greatest lower bound is $\sqrt{2}$.

The greatest lower bound is also called the *infimum*, and is written as $\inf\{x_\alpha; \alpha \in I\}$ or $\inf_{\alpha \in I} x_\alpha$. We similarly define the *supremum* of a set of real numbers to be their least upper bound. By convention, for the empty set $\emptyset$, we define $\inf \emptyset = \infty$ and $\sup \emptyset = -\infty$. Also, if a set S is *not* bounded below, then $\inf S = -\infty$; similarly, if it is not bounded above, then $\sup S = \infty$. One obvious but useful fact is

$$\inf S \;\leq\; x, \qquad \text{for any } x \in S. \tag{A.4.1}$$

Of course, if a set of real numbers has a minimal element (for example, if the set is finite), then this minimal element is the infimum. However, infimums exist even for sets without minimal elements. For example, if S is the set of all positive real numbers, then $\inf S = 0$, even though $0 \notin S$.

A simple but useful property of infimums is the following.

Proposition A.4.2. *Let S be a non-empty set of real numbers which is bounded below. Let $a = \inf S$. Then for any $\epsilon > 0$, there is $s \in S$ with $a \leq s < a + \epsilon$.*

Proof. Suppose, to the contrary, that there is no such s. Clearly there is no $s \in S$ with $s < a$ (otherwise a would not be a lower bound for S); hence, it must be that all $s \in S$ satisfy $s \geq a + \epsilon$. But in this case, $a + \epsilon$ is a lower bound for S which is larger than a. This contradicts the assertion that a was the *greatest* lower bound for S. ∎

For example, if S is the interval $(5, 20)$, then $\inf S = 5$, and $5 \notin S$, but for any $\epsilon > 0$, there is $x \in S$ with $x < 5 + \epsilon$.

Exercise A.4.3. (a) Compute $\inf\{x \in \mathbf{R} : x > 10\}$.
(b) Compute $\inf\{q \in \mathbf{Q} : q > 10\}$.
(c) Compute $\inf\{q \in \mathbf{Q} : q \geq 10\}$.

Exercise A.4.4. (a) Let $R, S \subseteq \mathbf{R}$ each be non-emtpy and bounded below. Prove that $\inf(R \cup S) = \min\left(\inf R,\ \inf S\right)$.
(b) Prove that this formula continues to hold if $R = \emptyset$ and/or $S = \emptyset$.
(c) State and prove a similar formula for $\sup(R \cup S)$.

Exercise A.4.5. Suppose $\{a_n\} \to a$. Prove that $\inf_n a_n \leq a \leq \sup_n a_n$. [Hint: Use proof by contradiction.]

Two special kinds of limits involve infimums and supremums, namely the *limit inferior* and the *limit superior*, defined by $\liminf_n x_n = \lim_{n \to \infty} \inf_{k \geq n} x_k$ and $\limsup_n x_n = \lim_{n \to \infty} \sup_{k \geq n} x_k$, respectively. Indeed, such limits always exist (though they may be infinite). Furthermore, $\lim_n x_n$ exists if and only if $\liminf_n x_n = \limsup_n x_n$.

Exercise A.4.6. Suppose $\{a_n\} \to a$ and $\{b_n\} \to b$, with $a < b$. Let $c_n = a_n$ for n odd, and $c_n = b_n$ for n even. Compute $\liminf_n c_n$ and $\limsup_n c_n$.

We shall sometimes use the "order" notations, $O(\cdot)$ and $o(\cdot)$. A quantity $g(x)$ is said to be $O(h(x))$ as $x \to \ell$ if $\limsup_{x \to \ell} |g(x)/h(x)| < \infty$. (Here ℓ can be ∞, or $-\infty$, or 0, or any other quantity. Also, $h(x)$ can be any function, such as x, or x^2, or $1/x$, or 1.) Similarly, $g(x)$ is said to be $o(h(x))$ as $x \to \ell$ if $\lim_{x \to \ell} g(x)/h(x) = 0$.

Exercise A.4.7. Prove that $\{a_n\} \to 0$ if and only if $a_n = o(1)$ as $n \to \infty$.

Finally, we note the following. We see by Exercise A.3.5(b) and induction that if $\{x_{nk}\}$ are real numbers for $n \in \mathbf{N}$ and $1 \le k \le K < \infty$, with $\lim_{n \to \infty} x_{nk} = 0$ for each k, then $\lim_{n \to \infty} \sum_{k=1}^{K} x_{nk} = 0$ as well. However, if there are an *infinite* number of different k then this is not true in general (for example, suppose $x_{nn} = 1$ but $x_{nk} = 0$ for $k \ne n$). Still, the following proposition gives a useful condition under which this conclusion holds.

Proposition A.4.8. *(The M-test.)* Let $\{x_{nk}\}_{n,k \in \mathbf{N}}$ *be a collection of real numbers. Suppose that* $\lim_{n \to \infty} x_{nk} = a_k$ *for each fixed* $k \in \mathbf{N}$. *Suppose further that* $\sum_{k=1}^{\infty} \sup_n |x_{nk}| < \infty$. *Then* $\lim_{n \to \infty} \sum_{k=1}^{\infty} x_{nk} = \sum_{k=1}^{\infty} a_k$.

Proof. The hypotheses imply that $\sum_{k=1}^{\infty} |a_k| < \infty$. Hence, by replacing x_{nk} by $x_{nk} - a_k$, it suffice to assume that $a_k = 0$ for all k.

Fix $\epsilon > 0$. Since $\sum_{k=1}^{\infty} \sup_n x_{nk} < \infty$, we can find $K \in \mathbf{N}$ such that $\sum_{k=K+1}^{\infty} \sup_n x_{nk} < \epsilon/2$. Since $\lim_{n \to \infty} x_{nk} = 0$, we can find (for $k = 1, 2, \ldots, K$) numbers N_k with $x_{nk} < \epsilon/2K$ for all $n \ge N_k$. Let $N = \max(N_1, \ldots, N_K)$. Then for $n \ge N$, we have $\sum_{k=1}^{\infty} x_{nk} < K \frac{\epsilon}{2K} + \frac{\epsilon}{2} = \epsilon$.

Hence, $\lim_{n \to \infty} \sum_{k=1}^{\infty} x_{nk} \le \sum_{k=1}^{\infty} a_k$. Similarly, $\lim_{n \to \infty} \sum_{k=1}^{\infty} x_{nk} \ge \sum_{k=1}^{\infty} a_k$. The result follows. ∎

If $\lim_{n \to \infty} x_{nk} = a_k$ for each fixed $k \in \mathbf{N}$, with $x_{nk} \ge 0$, but if we do *not* know that $\sum_{k=1}^{\infty} \sup_n x_{nk} < \infty$, then we still have

$$\lim_{n \to \infty} \sum_{k=1}^{\infty} x_{nk} \ge \sum_{k=1}^{\infty} a_k \,, \tag{A.4.9}$$

assuming this limit exists. Indeed, if not then we could find some finite $K \in \mathbf{N}$ with $\lim_{n \to \infty} \sum_{k=1}^{\infty} x_{nk} < \sum_{k=1}^{K} a_k$, contradicting the fact that $\lim_{n \to \infty} \sum_{k=1}^{\infty} x_{nk} \ge \lim_{n \to \infty} \sum_{k=1}^{K} x_{nk} = \sum_{k=1}^{K} a_k$.

A.5. Equivalence relations.

In one place in the notes (the proof of Proposition 1.2.6), the idea of an equivalence relation is used. Thus, we briefly review equivalence relations here.

A *relation* on a set S is a boolean function on $S \times S$; that is, given $x, y \in S$, either x is related to y (written $x \sim y$), or x is not related to y (written $x \not\sim y$). A relation is an *equivalence relation* if (a) it is *reflexive*, i.e. $x \sim x$ for all $x \in S$; (b) it is *symmetric*, i.e. $x \sim y$ whenever $y \sim x$; and (c) it is *transitive*, i.e. $x \sim z$ whenever $x \sim y$ and $y \sim z$.

Given an equivalence relation $\sim$ and an element $x \in S$, the *equivalence class* of x is the set of all $y \in S$ such that $y \sim x$. It is straightforward to verify that, if $\sim$ is an equivalence relation, then any pair of equivalence classes is either identical or disjoint. It follows that, given an equivalence relation, the collection of equivalence classes form a *partition* of the set S. This fact is used in the proof of Proposition 1.2.6 herein.

Exercise A.5.1. For each of the following relations on $S = \mathbf{Z}$, determine whether or not it is an equivalence relation; and if it is, then find the equivalence class of the element 1:

(a) $x \sim y$ if and only if $|y - x|$ is an integer multiple of 3.
(b) $x \sim y$ if and only if $|y - x| \leq 5$.
(c) $x \sim y$ if and only if $|x|, |y| \leq 5$.
(d) $x \sim y$ if and only if either $x = y$, or $|x|, |y| \leq 5$ (or both).
(e) $x \sim y$ if and only if $|x| = |y|$.
(f) $x \sim y$ if and only if $|x| \geq |y|$.

B. Bibliography.

The books in Subsection B.1 below should be used for background study, especially if your mathematical background is on the weak side for understanding this text. The books in Subsection B.2 are appropriate for lower-level probability courses, and provide lots of intuition, examples, computations, etc. (though they largely avoid the measure theory issues). The books in Subsection B.3 cover material similar to that of this text, though perhaps at a more advanced level; they also contain much additional material of interest. The books in Subsection B.4 do not discuss probability theory, but they do discuss measure theory in detail. Finally, the books in Subsections B.5 and B.6 discuss advanced topics in the theory of stochastic processes and of mathematical finance, respectively, and may be appropriate for further reading after mastering the content of this text.

B.1. Background in real analysis.

K.R. Davidson and A.P. Donsig (2002), Real analysis with real applications. Prentice Hall, Saddle River, NJ.

J.E. Marsden (1974), Elementary classical analysis. W.F. Freeman and Co., New York.

W. Rudin (1976), Principles of mathematical analysis (3rd ed.). McGraw-Hill, New York.

M. Spivak (1994), Calculus (3rd ed.). Publish or Perish, Houston, TX. (This is no ordinary calculus book!)

B.2. Undergraduate-level probability.

W. Feller (1968), An introduction to probability theory and its applications, Vol. I (3rd ed.). Wiley & Sons, New York.

G.R. Grimmett and D.R. Stirzaker (1992), Probability and random processes (2nd ed.). Oxford University Press.

D.G. Kelly (1994), Introduction to probability. Macmillan Publishing Co., New York.

J. Pitman (1993), Probability. Springer-Verlag, New York.

S. Ross (1994), A first course in probability (4th ed.). Macmillan Publishing Co., New York.

R.L. Scheaffer (1995), Introduction to probability and its applications (2nd ed.). Duxbury Press, New York.

B.3. Graduate-level probability.

P. Billingsley (1995), Probability and measure (3rd ed.). John Wiley & Sons, New York.

L. Breiman (1992), Probability. SIAM, Philadelphia.

K.L. Chung (1974), A course in probability theory (2nd ed.). Academic Press, New York.

R.M. Dudley (1989), Real analysis and probability. Wadsworth, Pacific Grove, CA.

R. Durrett (1996), Probability: Theory and examples (2nd ed.). Duxbury Press, New York.

W. Feller (1971), An introduction to probability theory and its applications, Vol. II (2nd ed.). Wiley & Sons, New York.

B. Fristedt and L. Gray (1997), A modern approach to probability theory. Birkhäuser, Boston.

J.C. Taylor (1997), An introduction to measure and probability. Springer, New York.

D. Williams (1991), Probability with martingales. Cambridge University Press.

B.4. Pure measure theory.

D.L. Cohn (1980), Measure theory. Birkhäuser, Boston.

J.L. Doob (1994), Measure theory. Springer-Verlag, New York.

G.B. Folland (1984), Real analysis: Modern techniques and their applications. John Wiley & Sons, New York.

P.R. Halmos (1964), Measure theory. Van Nostrand, Princeton, NJ.

H.L. Royden (1988), Real analysis (3rd ed.). Macmillan Publishing Co., New York.

W. Rudin (1987), Real and complex analysis (3rd ed.). McGraw-Hill,, New York.

B.5. Stochastic processes.

S. Asmussen (1987), Applied probability and queues. John Wiley & Sons, New York.

R. Bhattacharya and E.C. Waymire (1990), Stochastic processes with applications. John Wiley & Sons, New York.

J.L. Doob (1953), Stochastic processes (2nd ed.). John Wiley & Sons, New York.

D. Freedman (1983), Brownian motion and diffusion. Springer-Verlag, New York.

S. Karlin and H. M. Taylor (1975), A first course in stochastic processes. Academic Press, New York.

S.P. Meyn and R.L. Tweedie (1993), Markov chains and stochastic stability. Springer-Verlag, London.

E. Nummelin (1984), General irreducible Markov chains and non-negative operators. Cambridge University Press.

S. Resnick (1992), Adventures in stochastic processes. Birkhäuser, Boston.

B.6. Mathematical finance.

M. Baxter and A. Rennie (1996), Financial calculus: An introduction to derivative pricing. Cambridge University Press.

N.H. Bingham and R. Kiesel (1998), Risk-neutral valuation: Pricing and hedging of financial derivatives. Springer, London, New York.

R.J. Elliott and P.E. Kopp (1999), Mathematics of financial markets. Springer, New York.

I. Karatzas (1997), Lectures on the mathematics of finance. American Mathematical Society, Providence, Rhode Island.

I. Karatzas and S.E. Shreve (1998), Methods of mathematical finance. Springer, New York.

A.V. Mel'nikov (1999), Financial markets: Stochastic analysis and the pricing of derivative securities. American Mathematical Society, Providence, Rhode Island.

Index